MIX
Papier aus verantwortungsvollen Quellen
Paper from responsible sources
FSC® C105338

Chakole M.M.

# Heat Transfer Enhancement Techniques

With Special Attention to Passive Methods
of Heat Transfer Enhancement

Anchor Academic
Publishing

Chakole M.M.: Heat Transfer Enhancement Techniques. With Special Attention to Passive Methods of Heat Transfer Enhancement, Hamburg, Anchor Academic Publishing 2016

Buch-ISBN: 978-3-96067-049-0
PDF-eBook-ISBN: 978-3-96067-549-5
Druck/Herstellung: Anchor Academic Publishing, Hamburg, 2016

**Bibliografische Information der Deutschen Nationalbibliothek:**
Die Deutsche Nationalbibliothek verzeichnet diese Publikation in der Deutschen Nationalbibliografie; detaillierte bibliografische Daten sind im Internet über http://dnb.d-nb.de abrufbar.

**Bibliographical Information of the German National Library:**
The German National Library lists this publication in the German National Bibliography. Detailed bibliographic data can be found at: http://dnb.d-nb.de

All rights reserved. This publication may not be reproduced, stored in a retrieval system or transmitted, in any form or by any means, electronic, mechanical, photocopying, recording or otherwise, without the prior permission of the publishers.

Das Werk einschließlich aller seiner Teile ist urheberrechtlich geschützt. Jede Verwertung außerhalb der Grenzen des Urheberrechtsgesetzes ist ohne Zustimmung des Verlages unzulässig und strafbar. Dies gilt insbesondere für Vervielfältigungen, Übersetzungen, Mikroverfilmungen und die Einspeicherung und Bearbeitung in elektronischen Systemen.

Die Wiedergabe von Gebrauchsnamen, Handelsnamen, Warenbezeichnungen usw. in diesem Werk berechtigt auch ohne besondere Kennzeichnung nicht zu der Annahme, dass solche Namen im Sinne der Warenzeichen- und Markenschutz-Gesetzgebung als frei zu betrachten wären und daher von jedermann benutzt werden dürften.

Die Informationen in diesem Werk wurden mit Sorgfalt erarbeitet. Dennoch können Fehler nicht vollständig ausgeschlossen werden und die Diplomica Verlag GmbH, die Autoren oder Übersetzer übernehmen keine juristische Verantwortung oder irgendeine Haftung für evtl. verbliebene fehlerhafte Angaben und deren Folgen.

Alle Rechte vorbehalten

© Anchor Academic Publishing, Imprint der Diplomica Verlag GmbH
Hermannstal 119k, 22119 Hamburg
http://www.diplomica-verlag.de, Hamburg 2016
Printed in Germany

Dedicated to my parents

# TABLE OF CONTENTS

CHAPTER 1 – INTRODUCTION ..................................................................................... 7
    1.1   INTRODUCTION ........................................................................................... 7
    1.2   MOTIVATION ............................................................................................... 9
    1.3   OBJECTIVES ................................................................................................ 9
    1.4   PROPOSED WORK ..................................................................................... 10
    1.5   ORGANISATION OF REPORT .................................................................. 10

CHAPTER 2 – LITERATURE REVIEW ....................................................................... 11
    2.1   INTRODUCTION ......................................................................................... 11
    2.2   STATE OF ART OF REVIEW OF THE HEAT TRANSFER ENHANCEMENT BY TWISTED TAPE INSERTS WITHOUT NANOFLUID ...................... 11
    2.3   STATE OF ART OF REVIEW OF THE HEAT TRANSFER ENHANCEMENT BY TWISTED TAPE INSERTS ALONG WITH NANOFLUID ................. 14
    2.4   CONCLUDING REMARKS ........................................................................ 17
    2.5   CLOSURE ..................................................................................................... 17

CHAPTER 3 – THEORY AND DESIGN OF TUBULAR HEAT EXCHANGER WITH TWISTED TAPE INSERTS ............................................................................................. 18
    3.1   THEORY OF HEAT TRANSFER ENHANCEMENT TECHNIQUES ..... 18
    3.2   THEORY OF TWISTED TAPES ............................................................... 19
    3.3   CLASSIFICATION OF HEAT EXCHANGERS ....................................... 19
        3.3.1   Tubular Heat Exchanger ................................................................... 20
    3.4   DESIGN OF TUBULAR HEAT EXCHANGER ....................................... 21
        3.4.1   Assumption in Design of Tubular Heat Exchanger ......................... 21
        3.4.2   Design of Test Section ...................................................................... 21
    3.5   DUAL/TRIPLE/QUADRUPLE TWISTED TAPES .................................. 25

CHAPTER 4 – EXPERIMENTAL METHOD AND PREPRATION OF NANOFLUID ...... 27
    4.1   SPECIFICATION OF THE COMPONENTS OF EXPERIMENTAL SETUP ......... 27
        4.1.1   System Components .......................................................................... 28
        4.1.2   Measuring Instruments ...................................................................... 30
    4.2   EXPERIMENTAL SETUP .......................................................................... 31
    4.3   BASICS OF NANOFLUID ......................................................................... 33

- 4.3.1 Importance of Nanofluid in Heat Exchanger Application .................................. 34
- 4.3.2 Types of Nanoparticle and Base Fluid .............................................................. 34
- 4.3.3 Selection Criteria for $Al_2O_3$ Nanoparticle ......................................................... 35
- 4.3.4 Preparation of $Al_2O_3$ Nanofluid ....................................................................... 35
- 4.4 STABILITY OF NANOFLUID ................................................................................. 37
  - 4.4.1 Addition of Surfactant ...................................................................................... 38
  - 4.4.2 Overhead Stirring ............................................................................................. 38
  - 4.4.3 Ultra Sonication ................................................................................................ 38
  - 4.4.4 Stability Evaluation Methods ........................................................................... 39
- 4.5 PROPERTIES OF $Al_2O_3$ NANOFLUID .................................................................. 40
  - 4.5.1 Density .............................................................................................................. 40
  - 4.5.2 Specific Heat .................................................................................................... 40
  - 4.5.3 Thermal Conductivity ...................................................................................... 41
  - 4.5.4 Viscosity ........................................................................................................... 41
- 4.6 TUBULAR HEAT EXCHANGER TESTING AND MEASURED DATA ANALYSIS ............................................................................................................. 42
  - 4.6.1 Testing Conditions ........................................................................................... 42
  - 4.6.2 Measured Parameters ....................................................................................... 43
  - 4.6.3 Estimated Parameters ....................................................................................... 43

CHAPTER 5 – RESULTS AND DISCUSSIONS ................................................................. 44
- 5.1 VERIFICATION OF EXPERIMENTAL RESULTS ................................................. 44
- 5.2 EFFECT OF MULTIPLE TWISTED TAPES ............................................................ 47
  - 5.2.1 Heat Transfer .................................................................................................... 47
  - 5.2.2 Friction Loss ..................................................................................................... 48
  - 5.2.3 Thermal Performance Factor ............................................................................ 49
- 5.3 EFFECT OF CO/COUNTER TAPE ARRANGEMENT ........................................... 50
  - 5.3.1 Heat Transfer .................................................................................................... 50
  - 5.3.2 Friction Loss ..................................................................................................... 50
  - 5.3.3 Thermal Performance Factor ............................................................................ 50
- 5.4 EFFECT OF NANOFLUID CONCENTRATION ..................................................... 51
  - 5.4.1 Heat Transfer .................................................................................................... 51
  - 5.4.2 Friction Loss ..................................................................................................... 54
  - 5.4.3 Thermal Performance Factor ............................................................................ 58

| CHAPTER 6 – CONCLUSIONS AND FUTURE SCOPES | 61 |
|---|---|
| 6.1 CONCLUSIONS | 61 |
| 6.2 FUTURE SCOPE | 62 |
| REFERENCES | 64 |
| APPENDIX AND ANNEXURES | 68 |

# CHAPTER 1
# INTRODUCTION

This chapter presents an introduction to the background of this work throughout the course, which includes detailed introduction of the heat exchanger, twisted tape inserts and nanofluids. Then, the methods used for enhancement of heat exchanger performance is explained and proper method for heat transfer enhancement is suggested in terms of designing a multiple twisted tape inserts and experimentally investigating its performance. This work is often followed by the motivation required for this work, objectives set before carrying out the work. Finally, an organisation of the report is provided in brief.

## 1.1 INTRODUCTION

A heat exchanger is a device which is constructed to facilitate the heat transfer between one medium to another medium efficiently. The word "Exchanger" really applied to all types of equipment in which heat is exchanged but it is often used specially to the equipment in which heat is exchanged between two process streams that are at different temperature and are separated by a solid wall and where the two process fluid do not mix with each other. Heat exchanger is an important and expensive equipment that is used in almost all field of process such as food and dairy processes, waste heat recovery processes, air conditioning and refrigeration systems and also plants of oil, petrochemical, sugar, chemical reactors, pharmaceutical, power generation, etc. Energy recovery is the prime requirement of today to optimize the energy consumption in industry.

To achieve maximum utilization of thermal energy, several heat transfer enhancement techniques have been used in many thermal engineering applications such as nuclear reactor, chemical reactor, chemical process, automotive cooling, refrigeration, and heat exchanger, etc. Heat transfer enhancement techniques are powerful tools to increase heat transfer rate and thermal performance as well as to reduce the size of heat transfer system in installing and operating costs. Heat transfer enhancement in thermal systems can be carried out either by active or passive methods. In active methods, there is need of supplying external power source to the fluid or the equipment whereas in passive methods, heat transfer enhancement is done by turbulence promoters (such as special surface geometries, twisted tape, propeller, tangential inlet nozzle, snail entry, axial/radial guide vane, spiral fin) or fluid additives (such as nanofluid), without using any direct external power source. Due to easy installation/operation and cost saving, passive methods are extensively preferred for heat transfer enhancement.

One important group of devices used in passive method is swirl flow devices which produce secondary recirculation on the axial flow leading to an increase of tangential and radial turbulent fluctuation. This allows a greater mixing of fluid inside a heat exchanger tube and subsequently reduces a thickness of the boundary layer. Among the swirl generators of tube inserts, twisted tapes have gained great attention and widely used for producing compact heat exchangers and upgrading the heat transfer rate of the existing heat exchanger due to its low cost, acceptable thermal performance and ease of manufacture installation. Twisted tapes are generally equipped along the core tube to generate swirl causing the fluid transfer between the core tube and near wall tube. This leads to several mechanisms for heat transfer augmentation by improving flow velocities caused by partial blockage of the tube flow, which directs toward reducing the hydrodynamic or thermal boundary layer thickness. The hydraulic diameter reduction results in greater heat transfer coefficient, lengthening flow path in consequence of a helically twisting fluid motion, improving fluid mixing and thinning thermal boundary layer. However, more pumping power is required when twisted tapes are equipped inside the tube. Therefore, economic considerations has to be taken into account by using twisted tape with a proper geometry.

There are some earlier work, regarding twisted tape inserts, were performed by researcher. They have investigated effect of single twisted tape inserts on heat transfer enhancement for both laminar and turbulent flows. Apart from them very few researcher have worked on the multiple twisted tapes for heat transfer enhancement. Therefore, there is a large scope for doing research in the field of multiple twisted tapes with modifying different parameters and investigating the performance.

Now a days there is a trend to use nanofluid in heat exchanger to enhance the thermal performance. A nanofluid is a fluid prepared by dispersion of metallic or non-metallic nanoparticles or nanofibers with a typical size less than 100 nm in a liquid. Nanofluids have attracted huge interest because of their greatly enhanced thermal properties. Nanofluids are colloidal dilute dispersion of nanoparticles (generally less than 5% in volume) such as metals, oxides, carbides, or carbon nanotubes in conventional coolants or base fluid such as water, ethylene glycol, and oil. Miniaturization of electronic and other industrial component has led to the demand for development if new compact heat exchangers with higher rate of heat removal form cooling fluid. Hence multiple twisted tape inserts along with nanofluid fulfils this demand. Therefore, this work is also deals and decides the feasibility of use of multiple twisted tape inserts along with nanofluid in tubular heat exchanger.

## 1.2 MOTIVATION

A high cost of energy and material has resulted in an increased effort aimed at producing high performance heat exchanger equipment. The methods of improving convective heat transfer in the tubes of heat exchangers have been widely investigated by many researchers. Still there is a wide scope and challenging task for performance enhancement of heat exchanger. A lot of research in the area of improvement of design parameters such as fin thickness, fin spacing, tube diameter, tube spacing, coil width, etc. is being carried out for heat transfer enhancement from a quite long time.

The heat transfer enhancement techniques have been classified into two main categories such as active and passive. An active techniques which require external power for heat transfer augmentation, and passive technique needs no such external power for enhancement. One of the passive technique is use of twisted tape inserts in order to enhance the performance of heat exchanger, recently research is being started by use of nanofluid in heat exchangers. Therefore, in this work, use of nanofluid along with twisted tape inserts are being carried out.

## 1.3 OBJECTIVES

The main objective of this investigation is to study the performance of multiple twisted tape inserts with nanofluid in tubular heat exchanger. The proposed work includes the determination of

i. Overall heat transfer coefficient of water and nanofluid with/without multiple twisted tape inserts.
ii. The effect of single/dual/multiple twisted tape inserts on the overall heat transfer coefficient.
iii. The effect of single/dual/multiple twisted tape inserts on frictional pressure loss.
iv. Effect of Reynolds number on Thermal performance factor.
v. Effect of Nusselt number and Reynolds number on the heat transfer coefficient.
vi. Deciding the feasibility of multiple twisted tape inserts with and without nanofluid for enhancing the heat transfer rate.

## 1.4 PROPOSED WORK

It is proposed to study experimentally the effect of multiple twisted tape on heat transfer coefficient, pressure drop and thermal performance factor of tubular heat exchanger with water and $Al_2O_3$ nanofluid.

In this work, aluminium oxide as a nanoparticle and SLS (Sodium Lauryl Sulphate) as a surfactant is used in water for preparation of $Al_2O_3$ nanofluid. Three different volume percentage of aluminium oxide nanoparticles are added in the water such as 0.07%, 0.14% and 0.21%). Experimental investigations are carried out as per standard test procedure with constant heat flux condition at different mass flow rates.

## 1.5 ORGANISATION OF REPORT

Chapter 2 presents the background information on literature review related to proposed work which includes single and multiple twisted tape inserts along with and without nanofluid.

Chapter 3 presents the theory of heat transfer enhancement techniques, heat exchanger classification, basics of twisted tape inserts, design of tubular heat exchanger with various assumptions and the parameter considerations.

Chapter 4 deals with experimental work carried out for this work which includes different devices used in setup, test procedure adopted and preparation and stability of nanofluid, stability evaluation method, and various properties of nanofluid obtained from existing correlations. This chapter also includes testing and measured data analysis.

Chapter 5 includes result and discussions with the help of graphical representation. It includes comparison of experimental result obtained from various combinations of twisted tapes with water and various concentrations of nanofluid.

Chapter 6 concludes the present research study and also provides future scope.

# CHAPTER 2
# LITERATURE REVIEW

This chapter gives a reviews of the literature related to this work. It gives the survey of literature on twisted tape inserts and application of nanofluid in heat exchanger. This chapter also covers the concluding remarks on the summary of literature reviewed.

## 2.1 INTRODUCTION

Twisted tape inserts are the very cost effective method for heat transfer enhancement in heat exchanger. Hence, they have attracted considerable research attention. Many researcher have studied experimentally and numerically the performance of single twisted tapes for heat transfer enhancement. Very few researcher have studied the effect of multiple twisted tape on heat exchanger performance. Researcher have also attempted to investigate heat transfer enhancement in various type of heat exchanger by using new passive heat transfer enhancement techniques such as use of nanofluids. Paper reviewed for this work are categorized under the heat transfer enhancement by twisted tape inserts without nanofluid and heat transfer enhancement by twisted tape inserts along with nanofluid.

## 2.2 STATE OF ART OF REVIEW OF THE HEAT TRANSFER ENHANCEMENT BY TWISTED TAPE INSERTS WITHOUT NANOFLUID

**M.M.K. Bhuiya et al.** [1] studied experimentally, the influences of triple twisted tapes on heat transfer rate, friction factor and thermal enhancement efficiency. The investigations were conducted using the mild steel triple twisted tapes with four different twists under uniform heat flux condition. The experimental results demonstrated that the Nusselt number, friction factor and thermal enhancement efficiency increased with decreasing twist ratio. The results indicated that the presence of triple twisted tapes led to a higher increase in the heat transfer rate over the plain tube. Correlations were developed based on the data gathered during this work for predicting the heat transfer, friction factor and thermal enhancement efficiency through a circular tube fitted with triple twisted tape inserts in terms of twist ratio, Reynolds number and Prandtl number.

**Xiaoyu Zhang et al.** [2] numerical analysis, the heat and fluid-flows through a round tube fitted with triple or quadruple twisted tapes of different clearance, with the aim to verify the thought of core flow heat transfer enhancement and investigate the effect of multi longitudinal vortex on the flow, heat transfer and friction loss behaviour. The contour plots of predicted velocity, streamline and temperature are also presented. The obtained results show that, a maximum increase of 171% and 182% are observed in the Nusselt number by using triple and quadruple twisted tapes. And the friction factors of the tube fitted with triple and quadruple twisted tapes are around 4-7 times as that of the plain tube and the results verify the theory of the core flow heat transfer enhancement. Physical quantity synergy analysis is performed to investigate the mechanism of heat transfer enhancement.

**C. Thianpong et al.** [3] investigate experimentally, the influences of twin-counter/co-twisted tapes on heat transfer rate, friction factor and thermal enhancement index. The twin counter twisted tapes are used as counter-swirl flow generators while twin co-twisted tapes are used as co-swirl flow generators in a test section. The tests are conducted using the twin counter twisted tapes and twin co-twisted tapes with four different twists for different Reynolds numbers under uniform heat flux conditions. The experiments using the single twisted tape are also performed under similar operation test conditions, for comparison. The experimental results demonstrate that Nusselt number, friction factor and thermal enhancement index increase with decreasing twist ratio. The results also show that the twin counter twisted tapes are more efficient than the twin co-twisted tapes for heat transfer enhancement. In addition, the empirical correlations of the heat transfer, friction factor and thermal enhancement index are also reported.

**S. Eiamsa-ard et al.** [4] numerically analysis, the heat and fluid-flows through a round tube fitted with twisted tape, with the aim to investigate the effect of tape clearance ratio on the flow, heat transfer and friction loss behaviors. A finite volume method with the standard k–ε turbulence model, the Renormalized Group k–ε turbulence model, the standard k–ω turbulence model, and the Shear Stress Transport k–ω turbulence model, is used in the simulation. The computations show that predicted results by Shear Stress Transport k–ω turbulence, are in good agreement with the measurements than other models. The contour plots of predicted velocity vector, static pressure, temperature, and turbulent kinetic energy are also presented. The obtained results show that, the mean heat transfer rates for the tube with twisted tape inserts are higher than that for the plain tube. The thermal performance factor of the twisted tape is influenced by the clearance ratios and the best thermal performance factor at constant pumping power is found at zero clearance ratio i.e. tight-fit twisted tape.

**M.M.K. Bhuiya et al.** [5] explored the effects of the double counter twisted tapes on heat transfer and fluid friction characteristics in a heat exchanger tube. The double counter twisted tapes were used as counter-swirl flow generators in the test section. The experiments were performed with double counter twisted tapes of four different twist ratios (y = 1.95, 3.85, 5.92 and 7.75) using air as the testing fluid in a circular tube turbulent flow regime where the Reynolds number was varied from 6950 to 50,050. The experimental results demonstrated that the Nusselt number, friction factor and thermal enhancement efficiency were increased with decreasing twist ratio. The results also revealed that the heat transfer rate in the tube fitted with double counter twisted tape was significantly increased with corresponding increase in pressure drop. In the range of the present work, heat transfer rate and friction factor were obtained to be around 60 to 240% and 91 to 286% higher than those of the plain tube values, respectively. The maximum thermal enhancement efficiency of 1.34 was achieved by the use of double counter twisted tapes at constant blower power. In addition, the empirical correlations for the Nusselt number, friction factor and thermal enhancement efficiency were also developed, based on the experimental data.

**Halit Bas and Veysel Ozceyhan** [6] experimentally investigated the flow friction and heat transfer behaviour in a twisted tape swirl generator inserted tube. The twisted tapes are inserted separately from the tube wall. The effects of twist ratios ($y/D$ = 2, 2.5, 3, 3.5 and 4) and clearance ratios ($c/D$ = 0.0178 and 0.0357) are discussed in the range of Reynolds number from 5132 to 24,989, and the typical one ($c/D$ = 0) is also tested for comparison. Uniform heat flux is applied to the external surface of the tube wall. The air is selected as a working fluid. The obtained experimental results from the plain tube are validated by using well known equations given in literature. The using of twisted tapes supplies considerable increase on heat transfer and pressure drop when compared with those from the plain tube. The Nusselt number increases with the decrease of clearance ratio and twist ratio, also increase of Reynolds number. For all investigated cases, heat transfer enhancement tends to decrease with the increase of Reynolds number and to be nearly uniform for Reynolds number over 15,000 and $y/D$ lower than 3.0. The highest heat transfer enhancement is achieved as 1.756 for $c/D$ = 0.0178 and $y/D$ = 2 at Reynolds number of 5183. Consequently, the experimental results present that the best operating regime of all investigated twisted tape swirl generator inserts is detected at low Reynolds number, leading to more compact heat exchanger. The empirical correlations based on the experimental results of the present study are also given for prediction the heat transfer, friction factor and heat transfer enhancement.

## 2.3 STATE OF ART OF REVIEW OF THE HEAT TRANSFER ENHANCEMENT BY TWISTED TAPE INSERTS ALONG WITH NANOFLUID

**W.H. Azmi et al.** [7] have undertaken the experiments with $TiO_2$ for flow of nanofluid in a tube and with tape inserts for the determination of heat transfer coefficients and friction factor in the turbulent range of Reynolds numbers. The nanofluid heat transfer coefficients in the Reynolds number range of 8000 and 30,000 increased with volume concentration up to 1.0% and decreased at higher concentrations for flow in a tube. For flow over tape insert of twist ratio 5, the maximum enhancement in heat transfer coefficient is 81.1% compared to water and 48.1% with nanofluid at 1.0% concentration. A comparison of heat transfer enhancements for flow of nanofluid in a tube with twisted tape insert is made considering pressure drop with Advantage Ratio. For flow with twisted tape inserts, the Advantage Ratio increases with twist ratio for both water and nanofluid.

**E. Esmaeilzadeh et al.** [8] carried out an experimental study to investigate heat transfer and friction factor characteristics of $\Upsilon$-$Al_2O_3$/water nanofluid through circular tube with twisted tape inserts with various thicknesses at constant heat flux. The twist ratio of twisted tape remained constant while the thicknesses were changed through three values. The experiments were performed in laminar flow regime of Reynolds numbers. Results indicated that twisted tape inserts enhanced the average convective heat transfer coefficient, and also more the thickness of twisted tape is more the enhancement of convective heat transfer coefficient is. Also, the highest enhancement was achieved at maximum volume concentration. Results showed that nanofluids have better heat transfer performance when utilized with thicker twisted tapes. At the same time, the increase in twisted tape thickness leads to an increase in friction factor. In the end, the combined results of these two phenomena result in enhanced convective heat transfer coefficient and thermal performance.

**Smith Eiamsa-ard and Kunlanan Kiat kittipong** [9] investigate experimentally, the thermal performance characteristics in a heat exchanger tube by multiple twisted tapes in different arrangements and $TiO_2$ nanoparticles with different concentrations as the working fluid. The tube inserted the multiple twisted tapes showed superior thermal performance factor when compared with plain tube or the tube inserted a single twisted tape, due to continuous multiple swirling flow and multi-longitudinal vortices flow along the test tube. The higher number of twisted tape inserts led to an enhancement of thermal performance that resulted from increasing contact surface area, residence time, swirl intensity and fluid mixing with multi-longitudinal vortices flow. Moreover, arrangement of twisted tapes in counter current was superior energy saving devices for the practical use, particularly at low Reynolds number. This was especially the case for quadruple counter tapes in the cross directions where heat transfer enhancement with relatively low friction loss penalty was deserved. This arrangement gives highest thermal performance factor with $TiO_2$ nanoparticle as a working fluid than using pure water.

**M.T. Naik et al.** [10] investigated experimentally, the heat transfer and friction factor of CuO nanoparticles dispersed in water/propylene blend in a plain tube with and without twisted tape inserts. Considerable enhancement in the Nusselt number is observed with CuO nanofluids over the base fluids and heat transfer enhancement is linearly proportional to the nanoparticle volume concentration in the base fluid. The increase friction factor of Nanofluids over the base fluid is not significant in a plane tube. The use of twisted tape inserts in CuO nanofluids enhances the heat transfer coefficient with little increment of friction factor and transfer enhancement is proportional to the number of twists on inserts. Correlations are developed to predict Nusselt number and friction factor for the flow of CuO nanofluids in a tube with and without twisted tape inserts.

**S. Eiamsa-ard and K. Wongcharee** [11] investigated experimentally, the combined effects of nanofluids, dual twisted-tapes and a micro-fin tube on the heat transfer rate, friction factor and thermal performance factor characteristics. The authors conducted experiments using the micro-fin alone as well as the micro-fin equipped with a single twisted tape for comparison. The experimental results revealed that the heat transfer rate increased with increasing nanofluid concentration. At similar operating conditions, the micro-fin tube equipped with dual twisted-tapes consistently gave superior thermal performance factor to the one equipped with a single twisted-tape as well as the micro-fin tube alone. For all cases, thermal performance factors were apparently above unity. This indicates the beneficial effect for the energy saving by the uses of the combined techniques.

**L. SyamSundar et al.** [12] studied experimentally, the turbulent convective heat transfer and friction factor characteristics of magnetic $Fe_3O_4$ nanofluid flowing through a uniformly heated horizontal circular tube with and without twisted tape inserts. Experiments are conducted in the particle different volume concentrations, twisted tape inserts of different twist ratios and Reynolds number range of 3000 to 22000. Heat transfer and friction factor enhancement of $Fe_3O_4$ nanofluid in a plain tube with twisted tape insert is 51.88% and 1.231 times compared to water flowing in a plain tube under same Reynolds number. Generalized regression equation is presented for the estimation of Nusselt number and friction factor for both water and $Fe_3O_4$ nanofluid in a plain tube and with twisted tape inserts under turbulent flow condition.

**L. Syam Sundar and K.V. Sharma** [13] experimentally determined the thermo physical properties like thermal conductivity and viscosity of $Al_2O_3$ nanofluid at different volume concentrations and temperatures and validated. Convective heat transfer coefficient and friction factor data at various volume concentrations for flow in a plain tube and with twisted tape insert is determined experimentally for $Al_2O_3$ nanofluid. Experiments are conducted in the Reynolds number range of 10,000–22,000 with tapes of different twist ratios in the range of $0 < H/D < 8.3$. The heat transfer coefficient and friction factor of 0.5% volume concentration of $Al_2O_3$ nanofluid with twist ratio of five is 33.51% and 1.096 times respectively higher compared to flow of water in a tube. A generalized regression equation is developed for the estimation of Nusselt number and friction factor valid for both water and nanofluid in plain tube and with inserts under turbulent flow conditions.

**Y. Raja Sekhar et al.** [14] conducted heat transfer experiments in a pipe under low Reynolds number range using water and water based nanofluids. Heat transfer coefficient and friction factor for nanofluid in the flow path enhanced compared to water. The experimental data is compared with the data of literature and are found to be in good agreement. The increase in heat transfer coefficient in plain tube with use of nanofluids is greater by 8-12% compared to the flow of water in a plain tube. The nanofluid of 0.5% particle concentration is having highest friction factor compared to water. The Nusselt number and friction factor increases with increase of particle concentration. But, friction factor decreases with increase of Reynolds number of flow whereas the Nusselt number increases. Using nanofluid with a high heat exchange can help in reduce the size of the heat exchanger or without increasing the size of the heat exchanger efficiency of the system can be improved. Further, using twisted tapes and nanofluids in the pipe flows is advantageous since it is visible from the results that the energy gained with heat exchange is more than the energy spent on pumping power. Finally, it was concluded that heat transfer enhancement in a horizontal tube increases with Reynolds number of flow and nanoparticle concentration.

## 2.4 CONCLUDING REMARKS

The final remarks made from the literature review reveals that there is a large scope for investigating the performance of multiple twisted tape inserts along with nanofluid at various parameters which are summaries as,

1. A number of experimental studies have been reported to investigate the effects of various inserts for performance evaluation.
2. Among the reported research works on different types of inserts performed by the different researchers, it is clear that a few research works were presented on multiple twisted tape inserts.
3. Most of research works has done on enhancement of heat transfer by nanofluid but heat transfer enhancement by using twisted tape together with nanofluids are limited explored.
4. Experimental study of heat transfer enhancement by using multiple twisted tapes together with $Al_2O_3$ Nanofluid is not completely available in literature.

From above it is clear that number of experimental studies have been reported for nanofluids for enhancement of heat transfer. Heat transfer enhancement by inserting different swirl flow devices in flow path is main point of interest for most of research person in recent year. However, there is need to study experimentally the combine effect of multiple twisted tapes and nanofluid. Keeping this aspect in the mind nanofluid with the multiple twisted tape inserts in place of single twisted tape is been selected for this work.

## 2.5 CLOSURE

Based upon the experimental and numerical studies reported in preceding section, it is evident that lot of work is carried out in order to understand the reason of heat transfer enhancement by single twisted tape insert. Very little work is carried out on multiple twisted tape inserts along with nanofluid. Therefore, there is a lot of scope to investigate the performance of multiple twisted tape inserts along with nanofluid.

# CHAPTER 3
# THEORY AND DESIGN OF TUBULAR HEAT EXCHANGER WITH TWISTED TAPE INSERTS

This chapter covers the theory of heat transfer enhancement techniques, theory of twisted tapes, classification of heat exchangers and brief description of twisted tape inserts. Also this chapter includes the design procedure of tubular heat exchanger.

## 3.1 THEORY OF HEAT TRANSFER ENHANCEMENT TECHNIQUES

Several heat transfer enhancement (HTE) techniques have been used in many engineering applications such as nuclear reactor, chemical reactor, chemical process, automotive cooling, refrigeration, and heat exchanger, etc. HTE techniques are powerful tools to increase heat transfer rate and thermal performance as well as to reduce of the size of heat transfer system in installing and operating costs. However, the need for miniaturization of thermal equipment has shifted the focus to the development of new high performance fluids with thermal conductivities higher than those of the conventional liquids. These high performance fluids can contribute to the evolution of space-saving yet cost effective thermal equipment with higher competitiveness in the global market.

Heat transfer enhancements can be achieved through active and passive methods as suggested by Ahuja [15] and Bergles [16]. Active heat transfer enhancement is achieved by the application of external energy on the fluid. Passive enhancement is attained by increasing the fluid surface area, providing artificially roughed surface, by turbulence promoter (such as special surface geometries, twisted tape, propeller, tangential inlet nozzle, snail entry, axial/radial guide vane, spiral fin) or fluid additives (such as nanofluid), of the passive methods, the heat transfer enhancement with nanofluid is highly encouraging. Due to its easy installation/operation and cost saving, passive method has drawn great attention.

Generally speaking, tube flow can be divided into two parts [17] the boundary flow and the core flow. The boundary flow is a fluid region near the wall in the tube, beyond which in the tube the core flow is defined. Heat transfer enhanced tubes such as [18-21] spiral grooved tube, longitudinal troughed tube, corrugated tube, inner-finned tube, spiral-ribbed tube, micro-ribbed tube and so on, are mainly considered to effectively design and improve heat transfer surface in the boundary flow. Moreover, those improved surfaces dominate convective heat transfer between the fluid and the tube wall. Therefore, this kind of methods can be called surface-based heat transfer enhancement or heat transfer enhancement in the boundary flow. On the contrary, the heat transfer enhancement in the core flow can be called fluid-based heat transfer enhancement. The surface-based heat transfer enhancement is the common method to enhance heat transfer in the tube. While these measures are effective for heat transfer, however, intensifying fluid disturbance in the boundary flow will result in more dissipation of fluid momentum, and enlarging continuously extended surface will cause more frictional resistance and viscosity dissipation. Thus, the flow resistance will be increased by adopting these techniques. If the flow resistance is overlarge, the fluid velocity will become small, which may weaken convective heat transfer between the fluid and the surface.

## 3.2 THEORY OF TWISTED TAPES

One important group of devices used in passive method is swirl flow devices which produce secondary recirculation on the axial flow leading to an increase of tangential and radial turbulent fluctuation. This allows a greater mixing of fluid inside a heat exchanger tube and subsequently reduces a thickness of the boundary layer [22-27] .Among the swirl generators of tube inserts, twisted tapes have gained great attention and widely used for producing compact heat exchangers and upgrading the heat transfer rate of the existing heat exchanger due to its low cost, acceptable thermal performance and ease of manufacture installation [28] .Twisted tapes are generally equipped along the core tube to generate swirl causing the fluid transfer between the core tube and near wall tube. This leads to several mechanisms for heat transfer augmentation by improving flow velocities caused by partial blockage of the tube flow, which directs toward reducing the hydrodynamic or thermal boundary layer thickness. The hydraulic diameter reduction results in greater heat transfer coefficient, lengthening flow path in consequence of a helically twisting fluid motion, improving fluid mixing and thinning thermal boundary layer. However, more pumping power is required when twisted tapes are equipped inside the tube. Therefore, economic consideration has to be taken into account by using twisted tape with a proper geometry.

## 3.3 CLASSIFICATION OF HEAT EXCHANGERS

In order to meet the widely varying applications, several types of heat exchangers have been developed which are classified on the basics of nature of heat exchange process, relative direction of fluid motion, design and construction features and physical state of fluids.

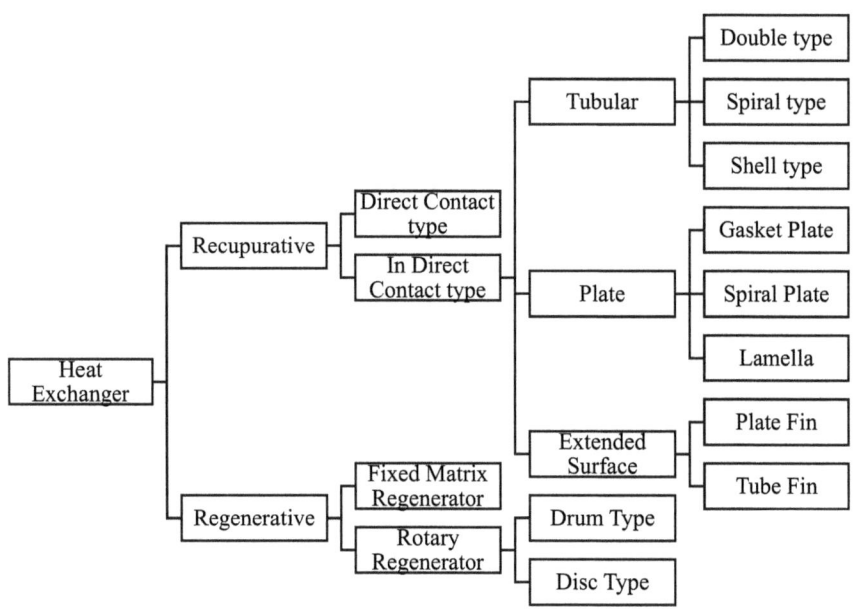

**Fig. 3.1 Classification of Heat Exchanger**

### 3.3.1 Tubular Heat Exchanger

Tubular heat exchangers are generally build of circular tubes, although elliptical, rectangular or round/flat twisted tubes. There is considerable flexibility in design of tubular heat exchanger because of core geometry can be varies easily by changing the tube diameter, length and arrangement. Tubular exchanger can be designed for high pressure relative to environment and high pressure differences between the fluids. Tubular exchangers are used primarily for liquid to liquid and liquid to phase change (Condensing and evaporating) heat transfer application. They are also used for gas to liquid and gas to gas heat transfer heat transfer application primarily when operating pressure and temperature is very high or fouling is a severe problem on at least one fluid side and no other types of exchanger can be used. These tubular heat exchangers may be classified as shell and tube type, double pipe type, and spiral tube type heat exchangers [29].

## 3.4 DESIGN OF TUBULAR HEAT EXCHANGER

The design of heat exchanger consist of the specification of the geometry (cross sectional area and length) that transfer the required heat load within the metallurgic limitation of material. Depending upon the heat exchanger design methodology, there are a set of geometric parameters that needs to be specified before the start of design.

### 3.4.1 Assumption in Design of Tubular Heat Exchanger
1. Properties of fluid are considered as constant, at an average value of inlet and outlet temperatures with little loss inaccuracy.
2. Constant heat flux boundary condition is considered.
3. Flow through heat exchanger is fully developed, steady and constant.
4. Fluid stream experiences little or no change in their velocities and elevations hence the kinetic energy and potential energy changes are negligible.
5. Outer surface of heat exchanger is assumed to perfectly insulated.
6. There is no fouling in heat exchanger.

### 3.4.2 Design of Test Section

The test section of this experimental setup consist of tube, heater and insulation as shown in fig. 3.2

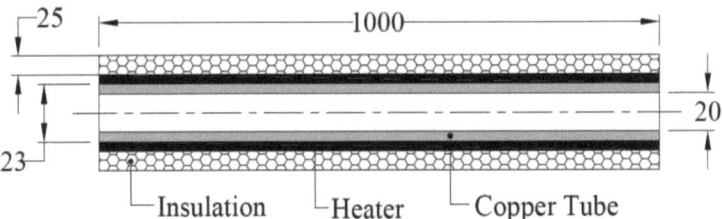

**Fig. 3.2 Schematic Diagram of Test Section**

#### 3.4.2.1 Tube Selection

Tube of test section is selected in such a way that, it provides minimum thermal resistance to flow of heat. For this purpose it is necessary to use material having high value of thermal conductivity. Among all engineering materials copper is most suitable one.

To get the best results from experiments flow through the tube must be hydro dynamically fully developed. For fully developed flow length of test section must be sufficiently larger. For this purpose 1 m long copper tube is chosen for experimentation. Test section must sustain pressure forces at higher temperature of fluid. Thickness of 3 mm is considered as safe value from previous literature. Specification of test section tube is given in table 3.1

**Table 3.1 Specifications of Test Section Tube**

| Material | Copper |
|---|---|
| Length | 1 m |
| Inner Diameter | 20 mm |
| Outer Diameter | 23 mm |

### 3.4.2.2 Heater Selection

Heater is to be wound on outer periphery of the test section tube. Heater is selected in such a way that it can provide constant heat flux condition with desired heat flux. To calculate the capacity of heater required, it is essential to know mass flow rate of fluid, heat transfer rate, and losses through test section. Minimum temperature difference of $1^0C$ is considered for heater design.

Mass flow rate through the test section is calculated as,

Assuming turbulent flow through test section with $R_e = 25000$,

$$R_e = \frac{\rho V D}{\mu}$$

$$25000 = \frac{(988.1)(V)(0.02)}{0.547 \times 10^{-3}}$$

$$V = 0.692 \text{ m/sec}$$

Mass flow rate:

$$\dot{m} = (\rho)(V)(A_c)$$

$$= (988.1)(0.692)(\frac{\pi}{4} 0.02^2)$$

$$= 0.859 \text{ kg/s}$$

Capacity of heater is calculated by,

Heat transfer rate

$$Q = \dot{m} C_p \Delta T$$

$$= (0.859)(4181.1)(1)$$

$$= 3.59 \text{ kW}$$

Assuming, losses of 4%

$$Q = (1.04)(3.59) = 3.74 \text{ kW} \cong 4 \text{ kW}$$

Hence at least heater of 4 kW capacity is required to fulfil the conditions of experiment. From the safety point of view, heater of capacity 6 kW is selected. Specification of heater used is given in table 3.2

Table 3.2 Specifications of Heater

| Heater type | Nichrome wire heater |
|---|---|
| Capacity | 6 kW |
| Diameter of wire | 1mm (20 Gauge) |
| Electric supply | 240V; 5A |
| Surface Temperature | $550^0C$ |

### 3.4.2.3 Insulation Selection

To avoid heat loss to the atmosphere it is necessary to cover test section with appropriate layer of insulation. Insulation should sustain temperature of heater i.e. $550^0C$ with minimum heat loss to atmosphere. Keeping this aspect in mind "Ceramic wool" is selected for insulation purpose.

Thickness of insulation must be selected from heat transfer point of view to achieve minimum heat loss to atmosphere. For proper selection of insulation, critical thickness of insulation is important to know, which is given by,

$$r_c = \frac{k}{h} = \frac{0.12}{10} = 0.012 \, m$$

$$t = r_c - r_o = 0.012 - 0.0115 = 0.005 \, m = 5mm$$

Any thickness of insulation more than 5mm will reduce heat loss. From heat transfer point of view, thickness of 25 mm is selected. Specification of insulation is given in table 3.3

**Table 3.3 Specifications of Insulation**

| Material | Ceramic wool |
|---|---|
| Max temperature sustain | $1260^0C$ |
| Density | 80-100 kg/m3 |
| Thermal conductivity | 0.12 W/m K |
| Thickness of insulation | 25 mm |

### 3.4.2.4 Pump Selection

Pump is selected in such a way that it should provide required mass flow rate of flow through test section. Power required, to pump fluid from water tank to different components of experimental setup is given by,

$$P_h = \rho\,Q\,g\,H$$

$$= (1000)\left(\frac{20}{60 X\ 1000}\right)(9.81)(50)$$

$$= 183.4\ W$$

Assuming pump efficiency η= 0.6

$$P = \frac{P_h}{\eta} = \frac{183.4}{0.6} = 305.67\ W$$

$$P = \frac{305.67}{0.746} = 0.41\ HP \cong 0.5 HP$$

Hence, a single stage centrifugal pump of capacity 0.5 HP is selected.

### 3.4.2.5 Selection of Sensors

To measure different parameters such as temperature, pressure drop, flow rate during experimentation, it is essential to install suitable sensors at required location. All sensors should have sufficient accuracy to predict performance of experimental setup.

Temperature sensor should have temperature range of $0-600^0C$. It should have resolution of minimum $0.1^0C$. With keeping this aspect in mind, K-Type thermocouple is selected. It has temperature range of $0-1260^0C$ with sensitivity of 40-55 $\mu V/^0C$ and a resolution of $0.1^0C$.

Due to turbulent flow in heat exchanger, flow meter should have range of 2-12 lpm. It should have least count of minimum 1 lpm. With keeping this aspect in mind, rotameter is selected for flow rate measurement. It has range of 2-20 lpm with a least count of 1 lpm.

Pressure sensor should be accurate enough to measure pressure drop across test section. With keeping this aspect in mind, digital type differential pressure transducer is selected. It has range of 0-9999.9 Pa with a least count of 0.1 Pa.

## 3.5 DUAL/TRIPLE/QUADRUPLE TWISTED TAPES

The schematic view and details of the single, dual, triple and quadruple twisted tapes with different arrangements are shown in fig. 3.3 and Table 3.4. All twisted tapes were made of aluminium strip with a thickness of 1 mm, which is a minimum twisting operation, and a length of 1000 mm. To fabricate a twisted tape, one end of a straight tape was clamped while another end was carefully twisted to achieve a desired twist length.

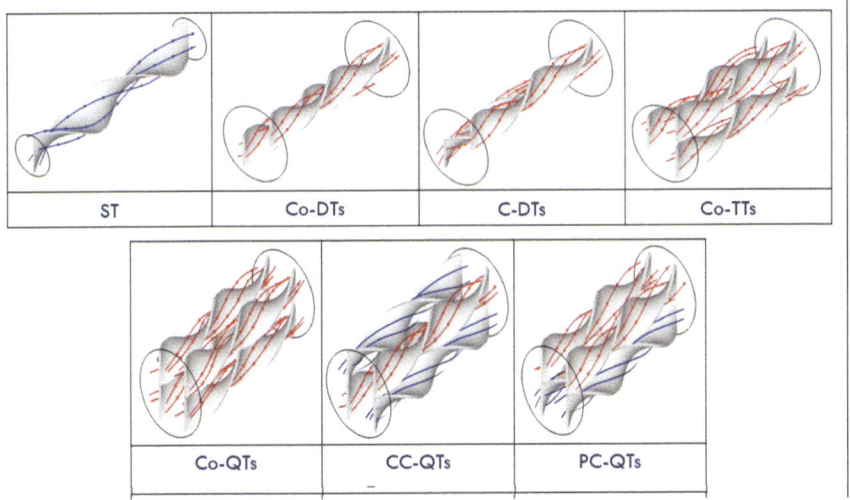

**Fig. 3.3 Configurations of Multiple Twisted Tapes**

Single twisted tape was 19 mm in width while dual, triple and quadruple twisted tapes were 8 mm in width. The tapes were formulated at constant twist ratio (y/W) of 5 where twist ratio is defined as twist length (180°/ twist length) to tape width (W). For dual, triple and quadruple twisted tapes, each tape was individually twisted and subsequently welded together. In the experiment, the swirl direction corresponding to tape arrangement was designed as: (i) co-swirl flow; all tapes were aligned to be twisted in the same direction. In this case, dual, triple and quadruple twisted tapes were assigned as Co-DTs/Co-TTs/Co-QTs, respectively, (ii) counter-swirl flow; this arrangement was designed for dual and quadruple twisted tapes. In the case of dual twisted tapes, two tapes were aligned to be twisted in opposite directions and assigned as C-DTs. In the case of quadruple twisted tapes, two tapes were aligned to be twisted in the same direction which was opposite to that of other two tapes. In addition, the quadruple counter tapes consisting of two pairs of tapes were in two different arrangements, to produce (1) parallel counter-swirl flow and (2) cross counter-swirl flow. For parallel counter-swirl flow, the tapes in each pair produced swirl flow in the same direction; in this case the quadruple counter tapes were assigned as PC-QTs. For cross counter-swirl flow, the tapes in each pair produced swirl flow in the opposite directions. The quadruple counter tapes were assigned as CC-QTs.

**Table 3.4 Configurations of Multiple Twisted Tapes**

| Twisted tape | ST | Co-DTs | C-DTs | Co-TTs | C-TTs | Co-QTs | PC-QTs | CC-QTs |
|---|---|---|---|---|---|---|---|---|
| (a) Number of tape | 1 | 2 | 2 | 3 | 3 | 4 | 4 | 4 |
| (b) Tape width (W) | 19 mm | 8 mm | 8 mm | 7.5 mm | 7.5 mm | 7 mm | 7 mm | 7 mm |
| (c) Tape pitch length (y) | 90 mm | 40 mm | 40 mm | 37.5 mm | 37.5 mm | 35 mm | 35 mm | 35 mm |
| (d) Twist ratio (y/W) | 5 | 5 | 5 | 5 | 5 | 5 | 5 | 5 |
| (e) Tape thickness ($\delta$) | 1 mm | 1 mm | 1 mm | 1 mm | 1 mm | 1 mm | 1 mm | 1 mm |
| (f) Material | Al | Al | Al | Al | Al | Al | Al | Al |
| (g) Swirl type | S–S | Co D-Ss | Counter D-Ss | Co T-Ss | Counter D-Ss | Co Q-Ss | Counter Q-Ss in P-A | Counter Q-Ss in C-A |

After completing theory, design and selection of components for complete experimental setup, basics and preparation methods of nanofluids along with experimental methods are discussed in neat chapter.

# CHAPTER 4
# EXPERIMENTAL METHOD AND PREPRATION OF NANOFLUID

This chapter covers the component used for fabrication of experimental setup and the test procedure adopted for experimentation. After that introduction and importance of nanofluid is discussed. Then the method adopted for preparation and stability of nanofluid is explained in detail followed by relations used for determining the thermo physical properties of nanofluid for each concentration.

## 4.1 SPECIFICATION OF THE COMPONENTS OF EXPERIMENTAL SETUP

For this work, the experimental setup fabricated is totally unique setup build for the investigation of performance of multiple twisted tape inserts. The different component of experimental setup are selected cautiously so that the component give the accurate performance under the application of nanofluid. The current section describes the specification of the experimental setup followed by the system component and details of the measuring instrument mounted on the setup. A specification of the present experimental setup is given in following table 4.1

**Table 4.1 Specification of the Experimental Setup**

| Sr. No. | Equipment Name | Dimension Range / Capacity | Remark |
|---|---|---|---|
| 1 | Test Section (Copper Tube) | Length= 1m, ID= 20mm, OD= 23mm | |
| 2 | Nichrome Wire Heater | 3 kW, 20 Gauge, 240V, 5A | Surface Temp=550$^0$C |
| 3 | Insulation (Ceramic wool) | Thickness=25 mm | K=0.12 W/mk |
| 4 | Pressure transducer | 0.1 to 9999.9 Pa | Least Count= 0.1 Pa |
| 5 | Pump | 0.5 HP | |
| 6 | Thermocouple (K Type) | 0$^0$C to 1260$^0$C | Least Count= 0.1$^0$C |
| 7 | Rotameter | 2 L/min to 20 L/min | Least Count= 0.5 L/min |
| 8 | Storage Tank | 25Litre | |

### 4.1.1 System Components

Experimental setup composed of following components which are explained below.

#### 4.1.1.1 Test Section

Test section consist of copper tube. Detailed specifications of copper tube is mentioned in table 4.1. Nichrome wire is wound on outer periphery of copper tube. Nichrome wire is properly insulated so that current will not pass to the copper tube and other components. Thermocouples are properly brazed at different positions on the periphery of copper tube. Pressure transducer is attached along the test section for measurement of pressure drop across test section. Copper tube along with Nichrome heater is thermally insulated to avoid heat loss to the environment. Test section is covered with metallic sheet to protect it from physical damage.

**Fig. 4.1 Photographic Image of Test Section**

#### 4.1.1.2 Twisted Tapes

All twisted tapes were made of aluminium strip with a thickness of 1 mm and a length of 1000 mm. To fabricate a twisted tape, one end of a straight tape was clamped while another end was carefully twisted to achieve a desired twist length.

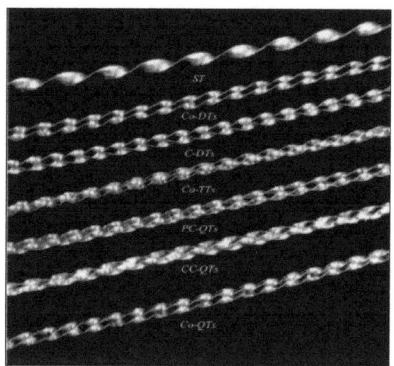

**Fig. 4.2 Photographic Image of Twisted Tapes**

Single twisted tape was 19 mm in width while dual, triple and quadruple twisted tapes were 8 mm, 7.5mm 7 mm in width respectively. For dual, triple and quadruple twisted tapes, each tape was individually twisted and subsequently welded together.

### 4.1.1.3 Evaporative Cooler

Evaporative cooler is used to achieve steady state conditions faster. It cools the hot liquid coming from the test section to its initial temperature in its operation, the hot water from the test section is pumped at the top of the cooler. Openings provided at the top sprays the hot liquid on the copper matrix. At the same time fan draws air from bottom side of the cooler and discharged out at the top of the cooler. This air causes the water from the surface of the copper matrix to evaporate and absorb the latent heat of evaporation from the remaining liquid. The air also absorb some sensible heat from liquid. Cold liquid collected at the bottom of cooler is directly supplied to the storage tank.

**Fig. 4.3 Photographic Image of Evaporative Cooler**

### 4.1.1.4 Pump

A single stage centrifugal pump of capacity 0.5 HP is used to lift and circulate water from the storage tank to test section. Inlet to the pump is taken from the storage tank and return line from the evaporative cooler is directly put into the storage tank. Outlet from the pump is directly connected to inlet at the test section.

Fig. 4.4 Photographic Image of Pump

### 4.1.2 Measuring Instruments

Experimental set up composed of many measuring and controlling devices which are explained below.

#### 4.1.2.1 Rotameters

There is only one Rotameter used in flow line. The Rotameter is installed just before the test section. Rotameter is an acrylic type Rotameter. It has a measuring range from 0 to 20 lpm with an accuracy of ±0.1 lpm.

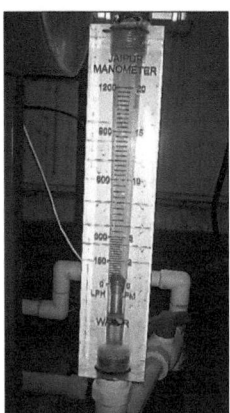

Fig 4.5 Photographic Image of Rotameter

### 4.1.2.2 Pressure Transducer

Pressure Transducer is used for measuring the pressure drop across inlet and outlet of test section. It has a measuring range from 0 to 9999 Pa with an accuracy of 0.1 Pa. Digital Pressure indicator directly shows difference in pressure across inlet and outlet of test section which is nothing but the pressure drop in test section.

**Fig 4.6 Photographic Image of Pressure Transducer**

### 4.1.2.3 Flow Control Bypass Valve

A manually operated flow control Bypass valve is installed in between pump and rotameter. Flow through the circuit can be controlled by this flow control valve. Excess fluid after flow control valve is drain back to the storage tank.

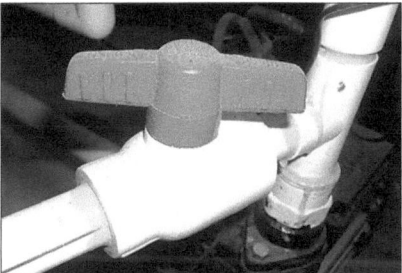

**Fig 4.7 Photographic Image of Flow control valve**

### 4.2 EXPERIMENTAL SETUP

The experimental setup consists of a test section, a evaporative cooler, a storage tank, and single stage centrifugal pump with a by-pass valve arrangement, temperature indicator, flow meter and a differential pressure transducer. The Schematic diagram of the experimental setup is shown in fig.4.8

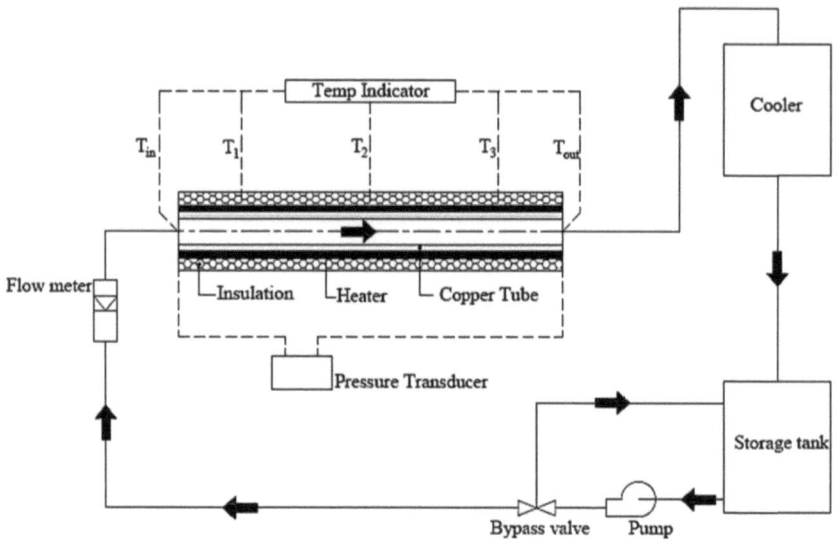

**Fig 4.8 Schematic Diagram of Experimental Setup**

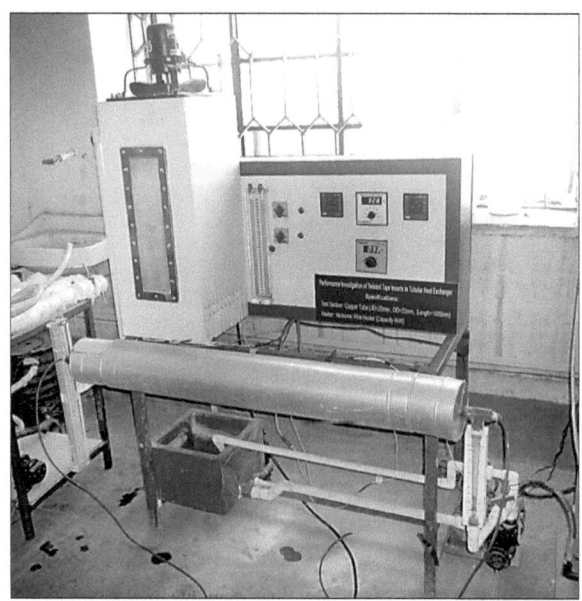

**Fig 4.9 Photographic Image of Experimental Setup**

Fig 4.9 shows the photographic image of experimental setup fabricated for this work. Test section consist of copper tube of 1m length having 20mm and 23mm inside and outside diameter respectively. Nichrome wire is wound on outer periphery of copper tube. Nichrome wire is properly insulated so that current will not pass to the copper tube and other components. Several thermocouples are provided, in which two are used to record the inlet, outlet temperatures of the working fluids and the remaining thermocouples are brazed on the outer periphery of the test section of the tube, to measure the average surface temperature of the tube. Pressure drop across the test section is measured by providing pressure transducer. Copper tube along with Nichrome heater is thermally insulated to avoid heat loss to the environment. Test section is covered with metallic sheet to protect it from physical damage. The aspect ratio of the test section is sufficiently large for the flow to be hydro-dynamically developed.

The working fluid under investigation is forced through the test section with pump connected to the sufficient capacity of storage tank. The fluid is heated by receiving heat from the test section and is allowed to cool by passing it through an evaporative cooler. By recirculation, the evaporative cooler in the flow loop helps in achieving steady state condition faster. As it is a closed system, all main components of experimental setup are connected with PVC pipes for flow circulation. A rotameter is incorporated in order to measure flow rate of working fluid.

Temperature indicator with $0.1^0C$ least count is used to indicate temperature readings of various thermocouples provided at different positions of test section. The physical properties of fluid flowing inside the tube of test section is assumed to be constant along the length and evaluated at the average bulk temperature for each run.

## 4.3 BASICS OF NANOFLUID

Nanofluid is the term first suggested by Choi, since then it gain the popularity. Nanofluids are produced by suspending Nano sized particles of high thermal conductivity metals and metal oxides, nanofibers in heat transfer liquids such as oil, water and ethylene glycol with at least one dimension less than 100 nanometres. Nanofluids have superior material properties than their base material and base fluid. Therefore, the addition of nanoparticles to the base fluid improves the thermal and fluid properties of nanofluid.

The thermo physical properties of nanofluids depend on its operating temperature. The important parameters which influence the heat transfer characteristics of nanofluids are its properties which include thermal conductivity, viscosity, specific heat and density. Nanofluids are used for heat transfer applications because they increase the heat transfer coefficient value. Nanofluids help in conserving heat energy.

A coolant, or heat transfer fluid, is used to prevent the overheating of equipment such as electronic devices and transportation. However, conventional heat transfer fluid such as water or ethylene glycol generally has poor thermal properties. So, many efforts of immersing small particles with high thermal conductivity in the liquid coolant have been done to enhance thermal properties of the conventional transfer fluids. The early research, which used suspension and dispersion of millimetre and micrometre sized particle, faced the major problem of poor suspension stability. Thus, a new class of fluid to improve both heat transfer performance and suspension stability is required in the industrial field. This motivation leads to the use of nanofluid for this work.

### 4.3.1 Importance of Nanofluid in Heat Exchanger Application

Nanofluid has higher thermal conductivity than base fluid and coolant fluid used in heat exchangers hence they increase the performance of heat exchanger. By the use of nanofluid in heat exchanger the overall heat transfer coefficient of heat exchanger increases. Nanofluid leads to the maximum recovery of heat from the heat exchanger. As the research is somewhat stagnant in the field of heat exchanger regarding performance enhancement by changing the geometry of heat exchanger and fins. Therefore, new alternative is searched for heat exchanger performance enhancement and that leads to the field of nanofluids.

Nanofluids possess immense potential of application to improve heat transfer and energy efficient in several areas including vehicular cooling in transportation, power generation, defence, nuclear, space, microelectronics and biomedical devices.

### 4.3.2 Types of Nanoparticle and Base Fluid

Nanofluids are new class of fluids engineered by dispersing nanometre sized materials i.e nanoparticles, nanofibers, nanotubes, nanowires, nanorods, nenosheet, or droplets in base fluids Materials commonly used as nanoparticles include chemically stable metals e.g gold, copper etc, metal oxides e.g. $Al_2O_3$, CuO, metal carbides e.g. SiC, carbon in various form e.g diamond, graphite and carbon nanotubes. Common base fluids are water, organic liquids e.g ethylene, tri-ethylene-glycols, refrigerants, oils and lubricant, bio-fluids, polymeric solutions and other common liquids.

### 4.3.3 Selection Criteria for $Al_2O_3$ Nanoparticle

1. The main reason behind the selection of the aluminium oxide for this work is that, $Al_2O_3$ has higher thermal conductivity.
2. From the literature reviewed for the work it has been concluded that $Al_2O_3$ nanoparticle has long term stability in the base fluid specifically in the distilled water as compared to the other types of base fluids.
3. The cost of the aluminium oxide nanoparticle is less as compared to that of the other nanomaterial.
4. Also from the literature it has been realized that the most of researchers used the $Al_2O_3$ nanoparticle for the enhancement of performance of shell and tube heat exchanger. Nanofluid does not cause problem of corrosion of copper tube of heat exchanger hence $Al_2O_3$ is selected for this work.

Table 4.2 shows the specifications of the $Al_2O_3$ Nanoparticles used for this work.

**Table 4.2 Specifications of $Al_2O_3$ Nanoparticles**

| Chemical Formula | Alpha $Al_2O_3$ |
|---|---|
| Colour | White |
| Morphological | Spherical |
| Specific surface area , $m^2/g$ | 90-160 |
| Density, g/cc | 3.69 |
| Average particle size, nm | 40-100 |

### 4.3.4 Preparation of $Al_2O_3$ Nanofluid

The amount of $Al_2O_3$ nanoparticles required for preparation of a particular volume concentration of nanofluids is calculated as

$$Mass\ fraction\ of\ Al_2O_3 = \frac{m_{np}}{m_{np}+m_{bf}} \qquad (4.1)$$

The powdered form of $Al_2O_3$ nanoparticle is shown in the figure 4.10

**Fig 4.10 Photographic Image of Powdered form of Al$_2$O$_3$ Nanoparticle**

The amount of Al$_2$O$_3$ nanoparticles required to prepare nanofluid of different (%) volume concentration in a 1 litres of water is summarized in the table 4.3 shown below

**Table 4.3 (%) Volume Concentrations with the Mass of Al$_2$O$_3$ Nanoparticle**

| Sr. No. | % Volume Concentration($\varnothing$) | Mass of Al$_2$O$_3$ ( in gram) in 1 Litre of Water |
|---|---|---|
| 1 | 0.07 | 2.77 |
| 2 | 0.14 | 5.52 |
| 3 | 0.21 | 8.27 |

Preparation of nanofluid is an important stage and nanofluid are prepared in a systematic and carefully. A stable nanofluid with uniform particle dispersion is required and the same nanofluid is used for experimental purpose. Basically 3 different methods are available for preparation of stable nanofluid and are discussed in details.

### A) Mixing of Nanopowder in the Base Fluid

In this method, the nanoparticles are directly mixed in the base fluid and thoroughly magnetic stirred. Nanofluids prepared by this method give poor suspension stability, because the nanoparticle settle down due to gravity, within a few minutes of nanofluid preparation.

The time of particle settlement depends on type of nanoparticles used, density and viscosity properties of the base fluids.

**B) Adding Surfactant to the Base Fluid**

In this method a small amount of suitable surfactant, generally one tenth of mass of nanoparticles is added to the base fluid and stirred continuously for few hours. Nanofluids prepared using surfactants will give a stable suspension with uniform particle dispersion in the base liquid. The nanoparticles remain in suspension state for a long period of time without settling down at the bottom of the container.

**C) Acid Treatment of Base Fluid**

The pH value of the base fluid can be lowered by adding a suitable acid to it. A stable nanofluid with uniform particle dispersion can be prepared by mixing nanoparticles in an acid treated base fluid. Acid treated nanofluids may cause corrosion of the pipe wall with prolonged usage of nanofluids. Hence acid treated base fluids are not preferred for preparation of nanofluids even though formation of stable nanofluids is possible when the application of the nanofluid is in heat exchanger.

After studying the above three method it has been clear that the only direct mixing of nanoparticles in base fluid has less stability and the Acid treated base fluid method may incur corrosion problem in heat exchanger pipes so the surfactant added method for preparation of nanofluid is selected for this work. (SLS) Sodium Lauryl sulphate is used as surfactant for the present work. The amount of nanoparticle required is derived from Eq.4.1 shown in table 4.3 for each (%) concentration is added into the distilled water for preparation of nanofluid.

## 4.4 STABILITY OF NANOFLUID

The purpose of this study is to investigate the settling and agglomeration of the nanofluid over the time period for its application in heat exchanger as a working fluid. After preparation of nanofluids, agglomeration might occur over the time which results in fast sedimentation of nanoparticles due to enhancement of downward body force. To provide better cooling effect using nanofluids in industry, they are expected to possess long term stability. During application of $Al_2O_3$, nanofluid in industry, stability might be one key issue. Therefore, this issue of fast sedimentation of $Al_2O_3$ in base fluid is to be resolved. During preparation of $Al_2O_3$ nanofluid these issues should be taken into account to make a stable nanofluid, which also has better thermo physical properties. There are many techniques for preparing the stable nanofluid. Few of them are used for this work which is discussed below.

### 4.4.1 Addition of Surfactant

Surfactants can be defined as chemical compounds added to nanoparticles in order to lower surface tension of liquids and increase immersion of particles. Several literatures talk about adding surfactant to nanoparticles to avoid fast sedimentation: however, correct quantity of surfactant should be added to the base fluid to achieve long term stability. For this work surfactant added is one tenth of mass of nanoparticles for every concentration of $Al_2O_3$ nanofluid.

### 4.4.2 Overhead Stirring

Overhead Stirring process is carried out on the each concentration of $Al_2O_3$ Nanofluid for preparing the stable $Al_2O_3$ nanofluid which is shown in fig. 4.11. This is done with the help of stirrer which is kept in the beaker containing $Al_2O_3$ nanofluid. The stirring is carried out at the speed of 1500 rpm for 1 hour for each beaker containing 1 litre of nanofluid.

**Fig. 4.11 Photographic Image of Magnetic Stirring**

### 4.4.3 Ultra Sonication

In the ultra-sonication process, the ultrasonic sound waves of 20 kHz are produced from the bottom of the container for about 1 hour for 1 litre of $Al_2O_3$ nanofluid. This will help to prepare a homogeneous mixture of $Al_2O_3$ nanofluid.

**Fig. 4.12 Photographic Image of Ultra sonication**

The ultra-sonication process is done at the atmospheric condition. The photographic image of ultra-sonication process of $Al_2O_3$ nanofluid is shown in fig. 4.12. No particle settlement was observed at the bottom of flask containing $Al_2O_3$ nanofluid even after nine hours. The $Al_2O_3$ nanofluid prepared is assumed to be an isentropic, Newtonian in behaviour and their thermo physical properties are uniform and constant throughout the experimentation.

Water + $Al_2O_3$     After Mixing     Non-Sonicated Mixer     Sonicated Mixer
(After 60 min)

**Fig 4.13 Photographic Image of $Al_2O_3$ Nanofluid after Preparation**

### 4.4.4 Stability Evaluation Methods

There are many methods developed for the evaluation of stability of nanofluids, few of them are used by many researchers for their experimental work out of which sedimentation photograph capturing method is the most simple and direct method which gives considerable accuracy for determining the stability of nanofluid over a time period, these common methods are listed as follows.

1. Sedimentation Photograph Capturing Method
2. Zeta Potential Test.
3. LUV-V is spectrophotometer.
4. TEM (Transmission Electron Microscopy) and SEM (Scanning Electron Microscopy)
5. 3∞ method.
6. Sedimentation balance method.

It has been introduced as a basic method for evaluating the stability of nanofluid. After preparation of nanofluid, it is kept in a stationary standing condition inside the beaker and settlement of particles was recorded continuously by capturing photos. Waiting time for capturing photographs depends upon the researchers. For instance, [30] takes the photographs of their samples within 24 hours after preparation. [31] Investigated sedimentation of $Al_2O_3$ inside water by capturing photo after 7 days of the sample nanofluid. But in this work, the photos capture after every two days from the day of preparation of nanofluid.

## 4.5 PROPERTIES OF $AL_2O_3$ NANOFLUID

The properties of $Al_2O_3$, nanofluids are different from the properties of conventional heat transfer fluids. For determining the thermal performance of heat exchanger various thermal properties of the fluid is required to be known. The properties of water are well known from the steam tables or from the literature different relations proposed by many researchers are used for finding the properties nanofluid. Detail procedure for determining few properties are discussed below.

### 4.5.1 Density

From the literature review it is found that the density obtained by the correlation is very nearer to the experimentally determined value. Hence for finding the density value of all the concentrations of $Al_2O_3$ nanofluid, the following correlation is used [32]. It is found that as the % concentration increases the density of $Al_2O_3$ nanofluid also increases.

$$\rho_{nf} = \emptyset \rho_p + (1 - \emptyset)\rho_w$$

### 4.5.2 Specific Heat

The specific heat is one of the important properties and plays an important role in influencing heat transfer rate of nanofluids. The specific heat of $Al_2O_3$ nanofluids decreases with increase in the concentration of nanofluids. Hence for finding the Specific heat of all the concentrations of $Al_2O_3$ nanofluid, following correlation is used [32]

$$C_{nf} = \frac{\emptyset(\rho\,C)_p + (1-\emptyset)(\rho\,C)_w}{\emptyset\rho_p + (1-\emptyset)\rho_w}$$

### 4.5.3 Thermal Conductivity

Experimental studies on nanofluids containing nanoparticles are expected to give more thermal conductivity and lower specific heats over conventional fluids. There are many reasons for higher thermal conductivity of $Al_2O_3$ nanofluid. Brownian motion of the particles and large surface area is one of the factors for enhanced thermal conductivity of nanofluids. Surface to volume ratio for nanoparticles is very high and this ratio increases with decrease in nanoparticle size. Probably this could also be one of the reasons for rise in the thermal conductivity of nanofluids. Thermal conductivity of nanofluid is obtained from the Maxwell model [33].

$$\frac{k_{nf}}{k_w} = \frac{k_p + 2k_w - 2\emptyset\,(k_w - k_p)}{k_p + 2k_w + \emptyset\,(k_w - k_p)}$$

### 4.5.4 Viscosity

From the literature review it is found that the viscosity value obtained by the correlation is very nearer to the experimentally determined value. Hence for finding viscosity value of all the concentration of nanofluid the following correlation is used [32]. It is found that as the (%) concentration increases the viscosity of nanofluid also increases.

$$\frac{\mu_{nf}}{\mu_w} = 1 + 2.5\,\emptyset$$

Table 4.4 shows the thermo physical properties of $Al_2O_3$ nanofluid for each concentration of $Al_2O_3$ nanofluid obtained by using above correlation at 300 K

**Table 4.4 Thermo-physical properties of $Al_2O_3$ Nanofluid**

| Sr. No. | (%) Volumetric concentration | µnf $\left(\frac{Ns}{m^2}\right)$ | ϱnf $\left(\frac{kg}{m^3}\right)$ | Cpnf $\left(\frac{J}{kg\,K}\right)$ | knf $\left(\frac{W}{mK}\right)$ |
|---|---|---|---|---|---|
| 1 | 0.07 | 1.005 | 1203.7 | 3397.9 | 0.745 |
| 2 | 0.14 | 1.154 | 1410.4 | 2845.8 | 0.897 |
| 3 | 0.21 | 1.304 | 1617.1 | 2434.8 | 1.075 |

## 4.6 TUBULAR HEAT EXCHANGER TESTING AND MEASURED DATA ANALYSIS

This section covers the description of the recommended standard testing conditions and actual test conditions maintained during experimentation work. Also it covers different measured parameters and estimated parameters.

### 4.6.1 Testing Conditions

The Bureau of Indian Standards has specified standard test conditions and test methods for almost all type of heat exchanger. It has also specified the special requirements of test rooms as well as accuracies of the instruments and gauges required to carry out the test.

Also, it is necessary to take some precautions while preparing test setup and while doing the actual tests. Some of these are given below.

1. All thermocouples attached to the tubes must be insulated by least 12.7 mm thick insulation.
2. The appliances must be tested with all its accessories in practically the same way as it will be installed in the field.
3. The measuring instruments must be properly maintained. Apart from periodic calibration, such cares as adjusting zero of voltmeter, ammeter, etc. are essential for accurate readings.
4. The test conditions must be maintained steady and the test must be run long enough to ensure that the thermal equilibrium is reached.
5. All thermocouples attached to the tubes must be insulated by at least 12.7 mm thick insulation.
6. The appliances must be tested with all its accessories in practically the same way as it will be installed in the field.
7. The measuring instruments must be properly maintained. Apart from periodic calibration, such cares as adjusting zero of voltmeter, ammeter, etc. are essential for accurate readings.
8. The test conditions must be maintained steady and the test must be run long enough to ensure that the thermal equilibrium is reached.

As actual testing is carried out at constant heat flux conditions only, in this work, there is deviation of the testing conditions that have been maintained, from the standard mentioned conditions. Actual conditions are somewhat different, due to the limitations of the test setup facility that was available.

### 4.6.2 Measured Parameters

In view of investigating system performance, following parameters are essential to be measured for a particular ambient and inside room conditions:

1. Mass flow rate of fluid
2. Inlet temperature of fluid
3. Outlet temperature of fluid
4. Surface temperature of test section
5. Pressure drop across test section

For water and different concentrations of nanofluid, all above measured parameters for different twisted tapes are listed in annexure I.

### 4.6.3 Estimated Parameters

In this analysis, from the measured parameters for water and different concentrations of nanofluid, following parameters are essential to be estimate. They are listed in annexure II.

1. Reynolds number
2. Nusselt number
3. Prandtl number
4. Nusselt number ratio
5. Friction factor
6. Friction factor ratio
7. Rate of heat transfer
8. Overall heat transfer coefficient
9. Thermal performance factor

Based on above experimental methods and preparation of nanofluid technique, experimentation is carried out and results obtained are discussed in the next chapter.

# CHAPTER 5
# RESULTS AND DISCUSSIONS

This Chapter covers the detail results and discussions of the experimentation conducted during this work. The various system affected parameters are analysed through the graphical representation. Each system affecting parameter is discussed based on the analysis of results obtained from the measured and estimated parameters.

Experimentation is carried out at constant heat flux condition. Different Configuration of twisted tapes are tested at different concentration of nanofluid. The performance investigation includes the calculation of overall heat transfer coefficient, Pressure drop, Nusselt Number, friction factor, Thermal performance factor at various mass flow rate.

## 5.1 VERIFICATION OF EXPERIMENTAL RESULTS
### A) Plain Tube without Twisted Tape

To gain confidence on experimental data throughout the research, the experimental data of the plain tube with water as the working fluid were firstly compared with those from the open literatures [32] which are Dittus–Boelter and Gnielinski correlations for Nusselt number and Blasius and Petukhov correlations for the friction factor as under.

Nusselt number correlations for the plain tube:

Correlation of Dittus-Boelter:

$$Nu = 0.023 Re^{0.8} Pr^{0.4}$$

Correlation of Gnielinski:

$$Nu = \frac{(f/8)(Re - 1000)Pr}{1 + 12.7(f/8)^{1/2}(Pr^{2/3} - 1)}$$

Friction factor correlation for the plain tube:

Correlation of Petukhov:

$$f = (0.79 \ln Re - 1.64)^{-2}$$

Correlation of Blasius

$$f = 0.318\, Re^{-0.25}$$

Verification of the heat transfer and friction factor in the plain tube is shown in fig. 5.1.

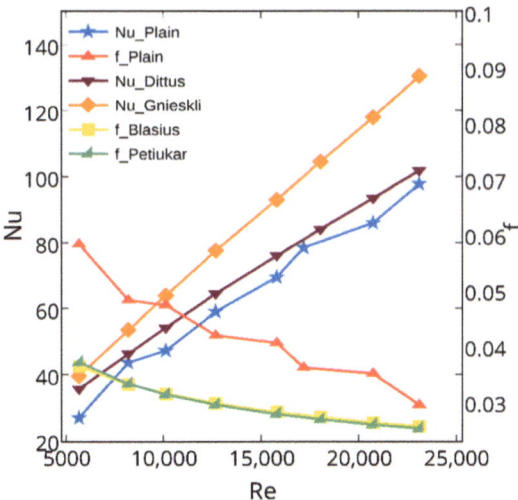

**Fig. 5.1 Reynolds Number variation with Nusselt Number of Plain tube**

The experimental Nusselt number was in satisfactory agreement, the mean experimental Nusselt number of plain tube were 5-12% and 18-26% lower than that of the Dittus-Boelter and Gnielinski correlations respectively. Experimental mean friction factor of plain tube were 33-50% and 34-51% higher than that of the Blasius and Petukhov correlations respectively. According to the comparative results mentioned above, it can be concluded that the present facility was reliable and experimental data was accurate enough. These provide a strong confidence in the present investigation of the heat transfer and flow friction in the tube.

**B) Plain Tube with Twisted Tape**

After getting a confidence on experimental data for plain tube without twisted tape inserts it is also essential to gain confidence on experimental data for plain tube with twisted tape inserts, the experimental data of the plain tube with water as the working fluid were compared with those from the open literatures [32] which are Manglic and Bergles correlations for Nusselt number and friction factor as under.

Nusselt number correlations for the typical twisted tape of Manglic and Bergles

$$Nu = \left[1 + \frac{0.769}{\left(\frac{y}{W}\right)}\right]\left[0.023 Re^{0.8} Pr^{0.4}\left(\frac{\pi}{\pi - \frac{4\delta}{D}}\right)^{0.8}\left(\frac{\pi + 2 - \frac{2\delta}{D}}{\pi - \frac{4\delta}{D}}\right)^{0.2}\right]\left(\frac{T_b}{T_w}\right)^{0.45}$$

Friction factor correlation for the typical twisted tape of Manglic and Borgles:

$$f = \left[1 + 2.06\left(1 + \left(\frac{2\left(\frac{y}{W}\right)}{\pi}\right)^2\right)^{-0.74}\right]\left[0.079Re^{-0.25}\left(\frac{\pi}{\pi - \frac{4\delta}{D}}\right)^{1.75}\left(\frac{\pi + 2 - \frac{2\delta}{D}}{\pi - \frac{4\delta}{D}}\right)^{1.25}\right]$$

Fig.5.2 shows the comparison between the experimental data of the present plain tube equipped with the typical/single twisted tape and those from the correlations of Manglik and Bergles [33].

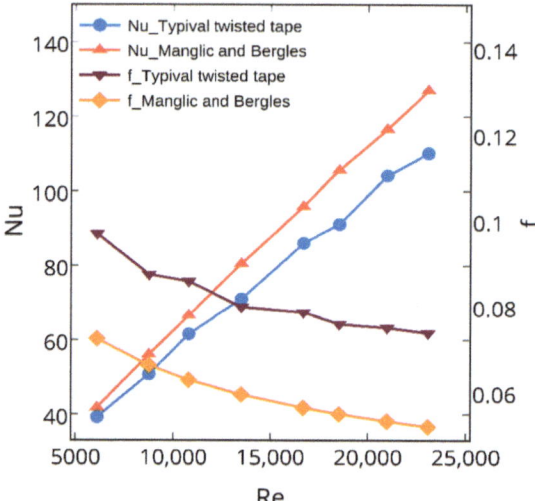

**Fig. 5.2 Reynolds Number variation with Nusselt Number of Typical Twisted tape**

Evidently, mean Nusselt number and friction factor of the tube with the typical/single twisted tape (ST) were 7-13% lower and 31-41% higher than those of Manglik and Bergles equation respectively. According to the comparative results mentioned above, it can be concluded that the present facility was reliable and experimental data was accurate enough. These provide a strong confidence in the present investigation of the heat transfer and flow friction in the tube equipped with dual/triple/quadruple twisted tapes.

## 5.2 EFFECT OF MULTIPLE TWISTED TAPES

### 5.2.1 Heat Transfer

The heat transfer of tubes equipped with tape insert(s) presented in terms of Nusselt number ($Nu_t$) and Nusselt number ratio ($Nu_t/Nu_p$), where $Nu_p$ is Nusselt number for the plain tube, is shown in fig. 5.3 and 5.4. For the present work, the Reynolds number is available over the range 5000–25,000 and water was used as the working fluid. fig. 5.3 shows that Nusselt number considerably increased with increasing Reynolds number. This was attributed to a stronger turbulent intensity and thus a better fluid mixing. At a given Reynolds number, Nusselt number in the tube with single/dual/triple/quadruple twisted tapes was significantly higher than those in the plain tube. This is responsible by the induction of multiple swirl flows, resulting in thinner boundary layer. fig. 5.4 shows that Nusselt number ratio ($Nu_t/Nu_p$) slightly decreased with increasing Reynolds number. This can be explained that at lower Reynolds number, a thermal boundary becomes thicker; therefore, the swirl flows induced by twisted tapes possess a more significant effect on disruption of thermal boundary. Moreover, a higher number of tape inserted in the tube consistently possessed higher Nusselt number. The tube with quadruple twisted tapes (QTs) provides the highest Nusselt number over the entire Reynolds number range. That is, Nusselt number of QTs was 21-31%, 44-56%, 69-81%, 100-132%, higher than those of the tubes with triple twisted tapes, dual twisted tapes, single twisted tape and the plain tube alone, respectively. This can be explained by the fact that more tapes induce higher number of swirl flows imparted to an axial flow, resulting in more uniform fluid mixing between the core and the tube wall regions, throughout the tube.

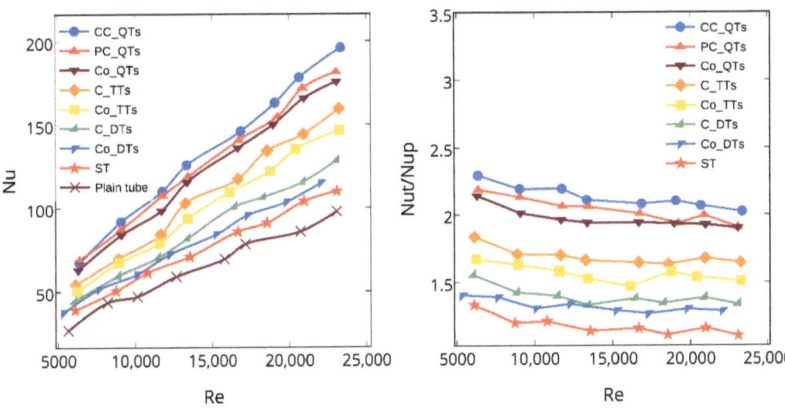

**Fig. 5.3 Reynolds Number variation with Nusselt Number of Different Tapes**

**Fig. 5.4 Reynolds Number variation with Nusselt Number Ratio of Different Tapes**

### 5.2.2 Friction Loss

The friction loss of tubes equipped with tape inserts presented in terms of friction factor ($f_t$) and friction factor ratio ($f_t/f_p$), where $f_p$ is friction factor for the plain tube, is shown in 5.5. For all the cases, friction factor slightly decreases. While friction factor ratio ($f_t/f_p$) slightly increases as shown in fig. 5.6 with increasing Reynolds number. The effect of twisted tape on friction factor was found that friction factors generated in the tube with tape insert(s) were considerably higher than those in the plain tube. In addition, multiple-tapes inserts (dual/triple/quadruple twisted tapes) consistently caused higher friction factor than the single one. This is directly responsible by the larger surface area of the inserts which perturbed the flows within the tubes. Moreover, an increase of swirl flow number boosted the interaction of the pressure forces with inertial forces in the boundary layer. Therefore, the highest mean friction factors were observed in quadruple twisted tapes. For co-swirl arrangement, friction factors over Co-QTs were 2.04 and 1.31 times higher than for Co-DTs (co-dual twisted tapes) and triple twisted tapes, respectively. For counter-swirl arrangement, CC-QTs generated 1.92 and 1.32 times more friction factors than C-DTs (counter-dual twisted tapes) and C-TTs respectively. The influence of multiple-tape inserts on friction factor ratio ($f_t/f_p$) was similar to that on friction factor ($f_t$). The CC-QTs gave the highest friction factor with the maximum friction factor ratio ($f_t/f_p$) of about 9.86.

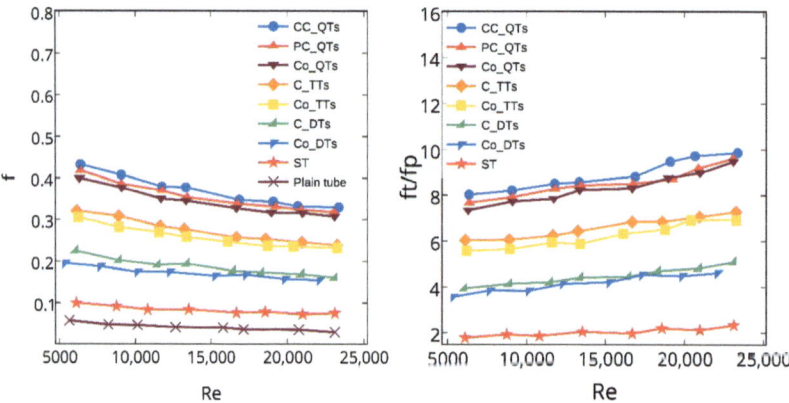

**Fig. 5.5 Reynolds Number variation with Friction factor of Different Tapes**

**Fig. 5.6 Reynolds Number variation with Friction factor Ratio of Different Tapes**

### 5.2.3 Thermal Performance Factor

Fig. 5.7 demonstrates the thermal performance factors ($\eta$) of tubes with tape inserts, which compromises between the heat transfer and friction loss. For a net energy gain, the value of the thermal performance factor is greater than unity. The results revealed that thermal performance factor decreases when Reynolds number increases. This implies the benefit of using tape inserts at lower Reynolds number rather than at higher Reynolds number. For a given Reynolds number, thermal performance factor increases as the number of tape inserts increases. It can be mentioned that the heat transfer enhancement by tape inserted reflects an overwhelming of increased friction loss. The thermal performance factors of the tubes with quadruple, triple and dual twisted tapes varied between 0.92-1.20, 0.86-1.05, and 0.79-0.99, respectively. In other words, quadruple and triple twisted tapes respectively gave 29-33%, 11-21% higher thermal performance factors than the single tape, whereas dual twisted tapes gave 6-12% higher thermal performance factors than the single tape.

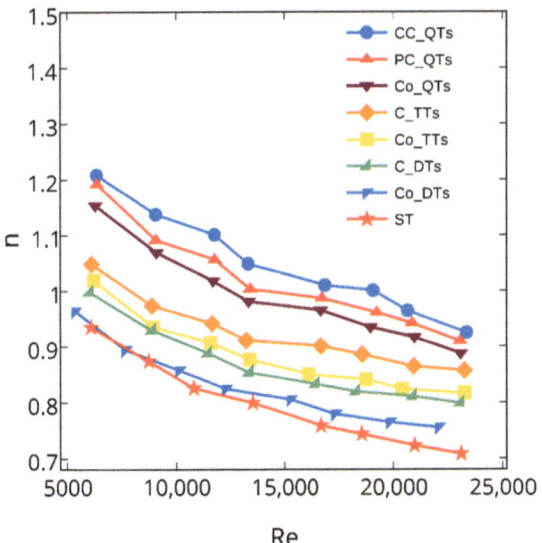

**Fig 5.7 Reynolds Number variation with Thermal Performance Factor of Different Twisted Tapes**

## 5.3 EFFECT OF CO/COUNTER TAPE ARRANGEMENT
### 5.3.1 Heat Transfer

The effect of co/counter tape arrangement on heat transfer enhancement can be observed from fig. 5.3 and fig. 5.4 The tapes in counter arrangement consistently yielded higher Nusselt number than that in co-arrangement.

$$Nu_{(C-DT)} > (11.3 - 17.6)\% \ of \ Nu_{(Co-DT)}$$
$$Nu_{(C-TT)} > (8.3 - 13.5)\% \ of \ Nu_{(Co-TT)}$$
$$Nu_{(CC-QT)} > (7.1 - 11.9)\% \ of \ Nu_{(Co-QT)}$$
$$Nu_{(PC-QT)} > (3.1 - 9.6)\% \ of \ Nu_{(Co-QT)}$$
$$Nu_{(PC-QT)} > (2.2 - 5.5)\% \ of \ Nu_{(PC-QT)}$$

### 5.3.2 Friction Loss

The effect of co/counter tape arrangement on friction loss is shown fig. 5.5 and fig. 5.6. For dual twisted tapes, C-DTs consistently caused higher friction factor than Co-DTs due to a higher Nusselt number reflecting to the pressure forces, as similarly.

$$Nu_{(C-DT)} > (4.3 - 11.5)\% \ of \ Nu_{(Co-DT)}$$
$$Nu_{(C-TT)} > (4.1 - 9.9)\% \ of \ Nu_{(Co-TT)}$$
$$Nu_{(CC-QT)} > (7.3 - 9.4)\% \ of \ Nu_{(Co-QT)}$$
$$Nu_{(PC-QT)} > (2.5 - 5.7)\% \ of \ Nu_{(Co-QT)}$$
$$Nu_{(PC-QT)} > (2.6 - 6.8)\% \ of \ Nu_{(PC-QT)}$$

### 5.3.3 Thermal Performance Factor

The effect of co-/counter tape arrangement on thermal performance factor at the same pumping power is shown in fig. 5.7. Apparently, the tapes in counter arrangement possessed higher thermal performance factors than that in co-arrangement. Therefore, it can be mentioned that in a case of DTs the heat transfer enhancement by C-DTs reflects an overwhelming of increased friction loss. With a tubes equipped with QTs, the highest thermal performance of 1.21 was observed for CC-QTs due to their excellent heat transfer enhancement with relatively low friction loss penalty. This highlights the important role of CC-QTs in improving the performance. In addition, it can be noted that the counter tape arrangement in all cases was superior energy saving devices for the practical use, particularly at low Reynolds number.

## 5.4 EFFECT OF NANOFLUID CONCENTRATION

### 5.4.1 Heat Transfer

Fig. 5.8 and Fig.5.9 presents heat transfer results of the $Al_2O_3$/water nanofluids (0.07%, 0.14% and 0.21% by volume) in accompanies with multiple twisted tapes. It was found that heat transfer (Nusselt number) increased when increasing nanofluid concentration. For the tubes with twisted tapes, the average Nusselt number of nanofluid with Al2O3 concentrations of 0.07%, 0.14% and 0.21% by volume, was around 3.3-7.4%, 9.8-17.9% and 18.6-24.6% those of the base fluid (water). In other words, nanofluid with $Al_2O_3$ concentration of 0.21% by volume gave 10.4-18.1% and 5.1-12.6% higher Nusselt number than the ones with $Al_2O_3$ concentrations of 0.07% and 0.14% by volume, respectively. Moreover, the increase of nanoparticle loading also offers higher contact area between nanoparticles and the base fluid as well as twisted tapes. The enhanced heat transfer is due to thermal conductivity and collision of nanoparticles.

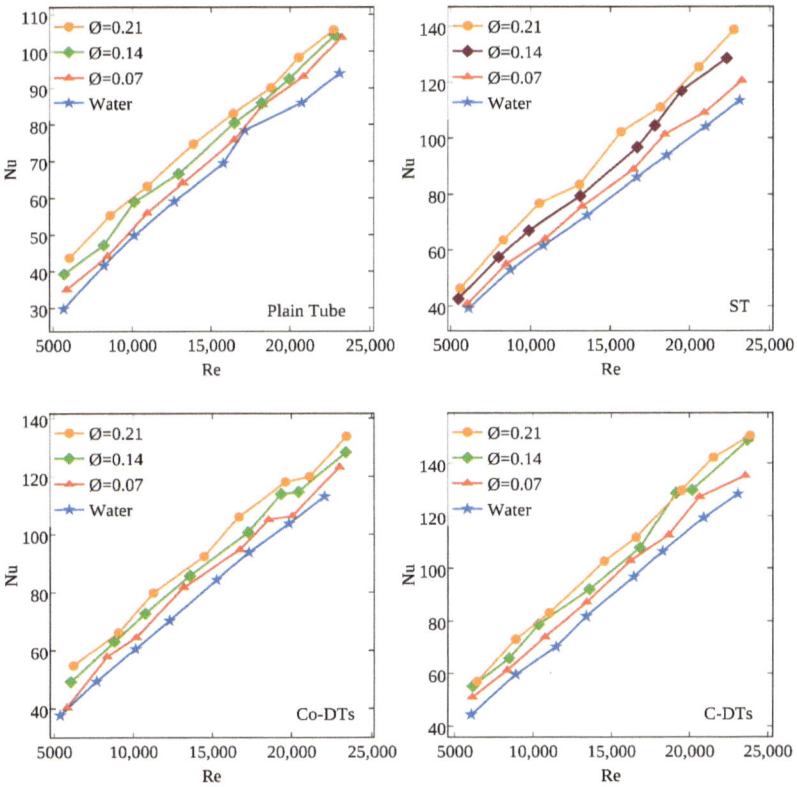

**Fig. 5.8 Reynolds Number variation with Nusselt number for Nanofluid**

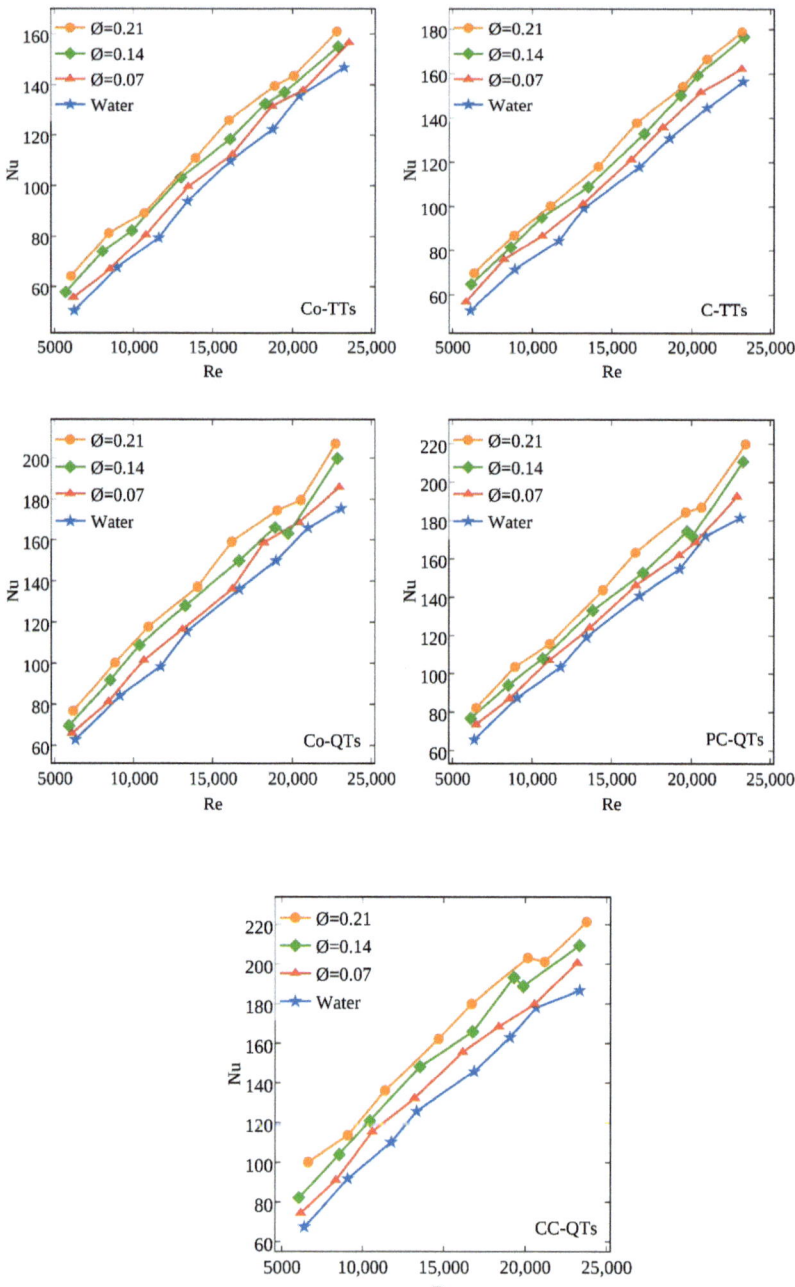

**Fig. 5.8 Reynolds Number variation with Nusselt number for Nanofluid (cont.)**

In particular, using CC-QTs induced an effective swirl flows in enhancing nanofluid exchange between the core and the wall regions and thus increasing convective heat transfer. As nanofluid with $Al_2O_3$ concentration was used, Nusselt number of CC-QTs was 7.5-17.6% and 12.0-18.2% higher than that of PC-QTs and Co-QTs, respectively. However, loading too much nanoparticles into nanofluid beyond the optimum concentration may diminish the fluid movement and heat transfer rate due to increased fluid viscosity. In the present work, the decrease of Nusselt number was not found when increasing in concentration. This implies the present nanofluid concentration range did not exceed the optimum level. Therefore, the effect of $Al_2O_3$ nanoparticles was more pronounced for thermal conductivity and the collision than viscosity.

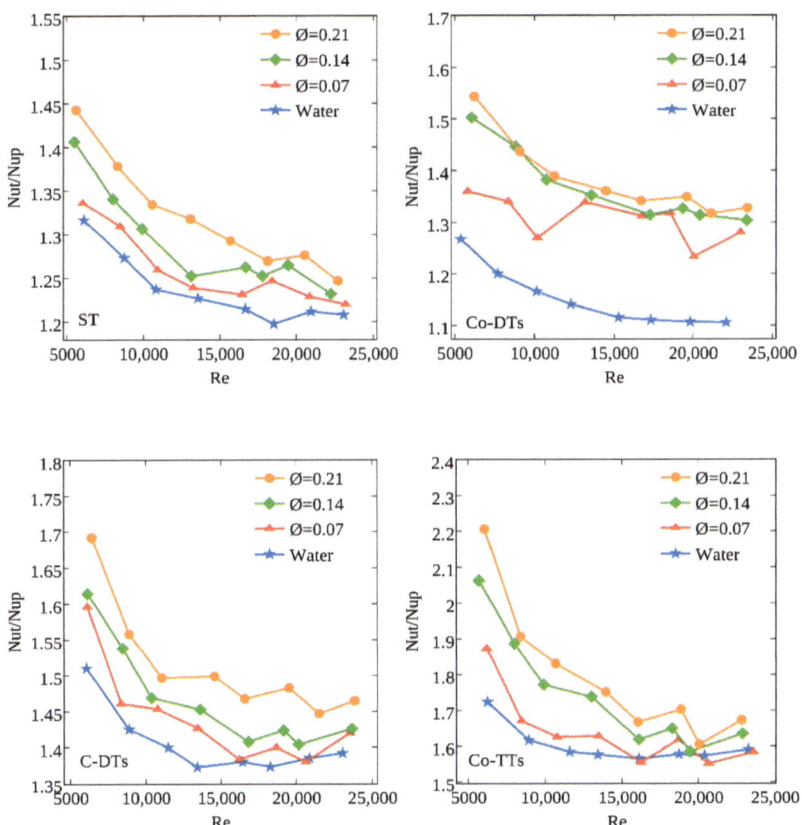

**Fig. 5.9 Reynolds Number variation with Nusselt number Ratio for Nanofluid**

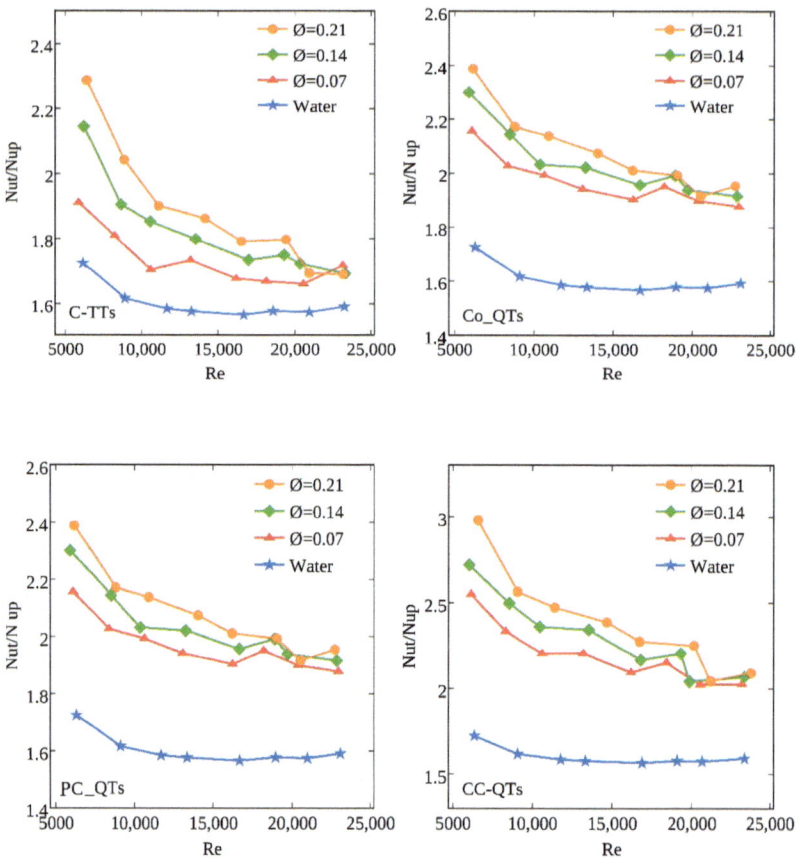

**Fig. 5.9 Reynolds Number variation with Nusselt number Ratio for Nanofluid (cont.)**

### 5.4.2 Friction Loss

The effect of nanofluid concentration on friction factor is shown in fig. 5.10. and fig.5.11 For the present range, nanofluid with $Al_2O_3$ concentrations of 0.07%, 0.14% and 0.21% by volume respectively caused 1.0-7.4%, 9.8-18.5% and 18.1-24.6% higher friction factor compared to those of the base fluid.

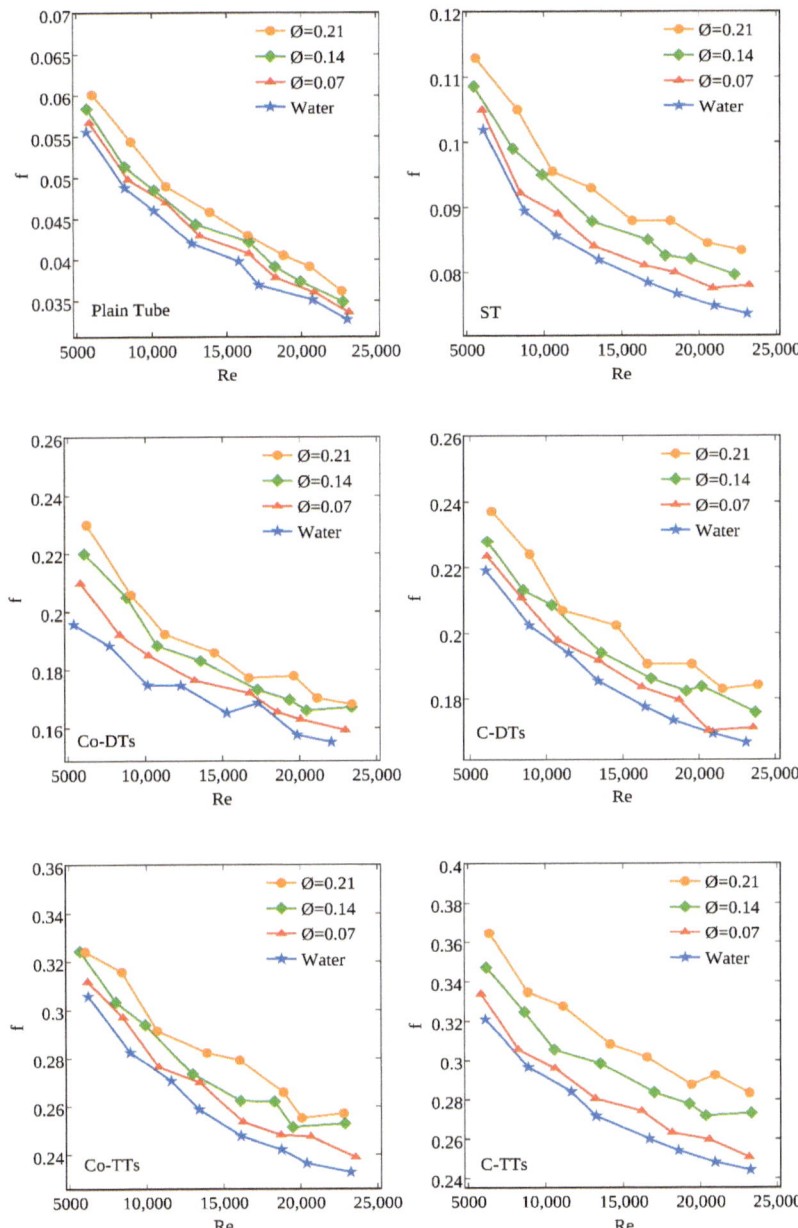

**Fig. 5.10 Reynolds Number variation with Friction Factor for Nanofluid**

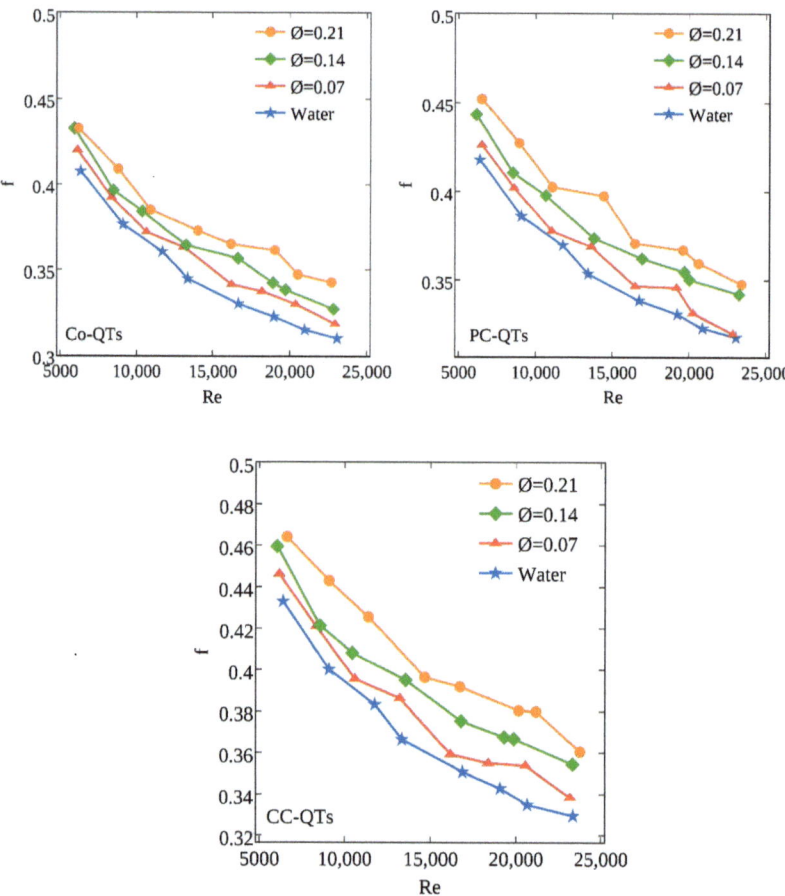

**Fig. 5.10 Reynolds Number variation with Friction Factor for Nanofluid (cont.)**

In other words, nanofluid with $Al_2O_3$ concentration of 0.21% by volume caused 2.6-7.5% and 1.0-5.1% higher friction factors than the ones with $Al_2O_3$ concentrations of 0.07% and 0.14% by volume, respectively. As nanofluid with $Al_2O_3$ concentration was used, the friction factors of Co-QTs were 2.6–5.6% and 5.2-8.3% higher than those of PC-QTs and CC-QTs, respectively. The increase of friction loss is directly caused by the increases of fluid viscosity and shear force on tube wall acted by nanoparticles. In particular, at low Reynolds number all nanofluid yielded higher pressure loss than the base fluid. However, the results indicate that utilizing nanofluid in the present concentration range is an insignificant friction loss penalty.

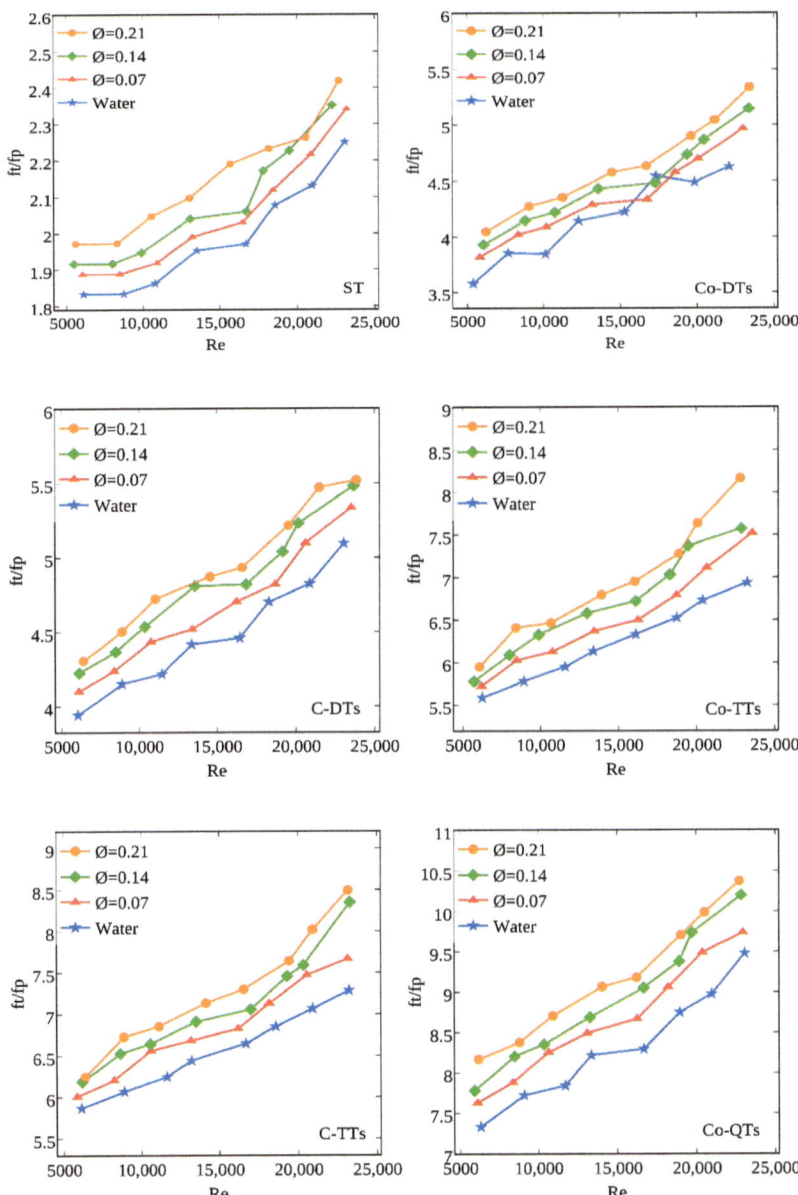

**Fig. 5.11 Reynolds Number variation with Friction Factor Ratio for Nanofluid**

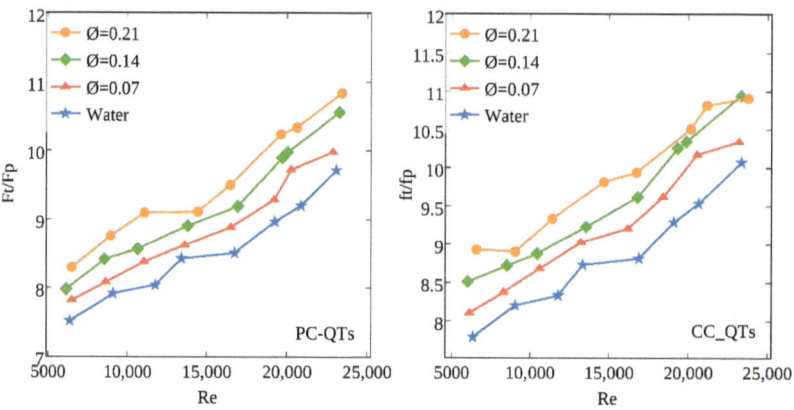

**Fig. 5.11 Reynolds Number variation with Friction Factor Ratio for Nanofluid (cont.)**

### 5.4.3 Thermal Performance Factor

Fig. 5.12 shows the effect of nanofluid concentration on thermal performance factor. Evidently, nanofluid with higher $Al_2O_3$ concentrations yielded higher thermal performance factors. Depending on Reynolds number, thermal performance factors given by nanofluid with $Al_2O_3$ concentrations of 0.07%, 0.14% and 0.21% by volume were 0.97-1.28, 1.06-1.46 and 1.04-1.64 respectively. Comparatively, nanofluid with $Al_2O_3$ concentration of 0.21% by volume offered 6.4-28.7% and 4.2-17.2% higher thermal performance factors than the ones with $Al_2O_3$ concentrations of 0.07% and 0.14% by volume, respectively.

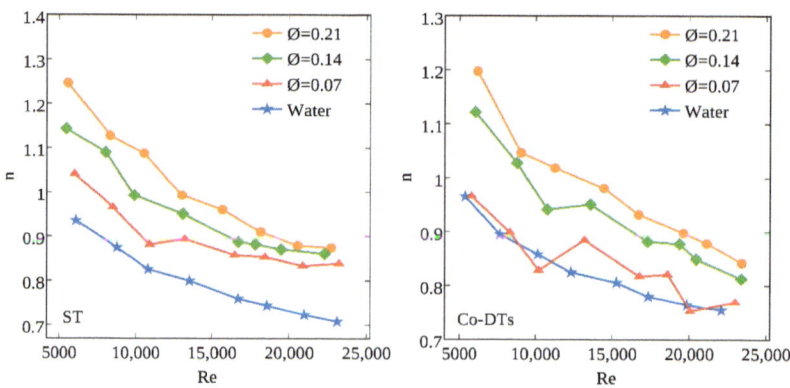

**Fig. 5.12 Reynolds Number variation with Thermal Performance Factor for Nanofluid**

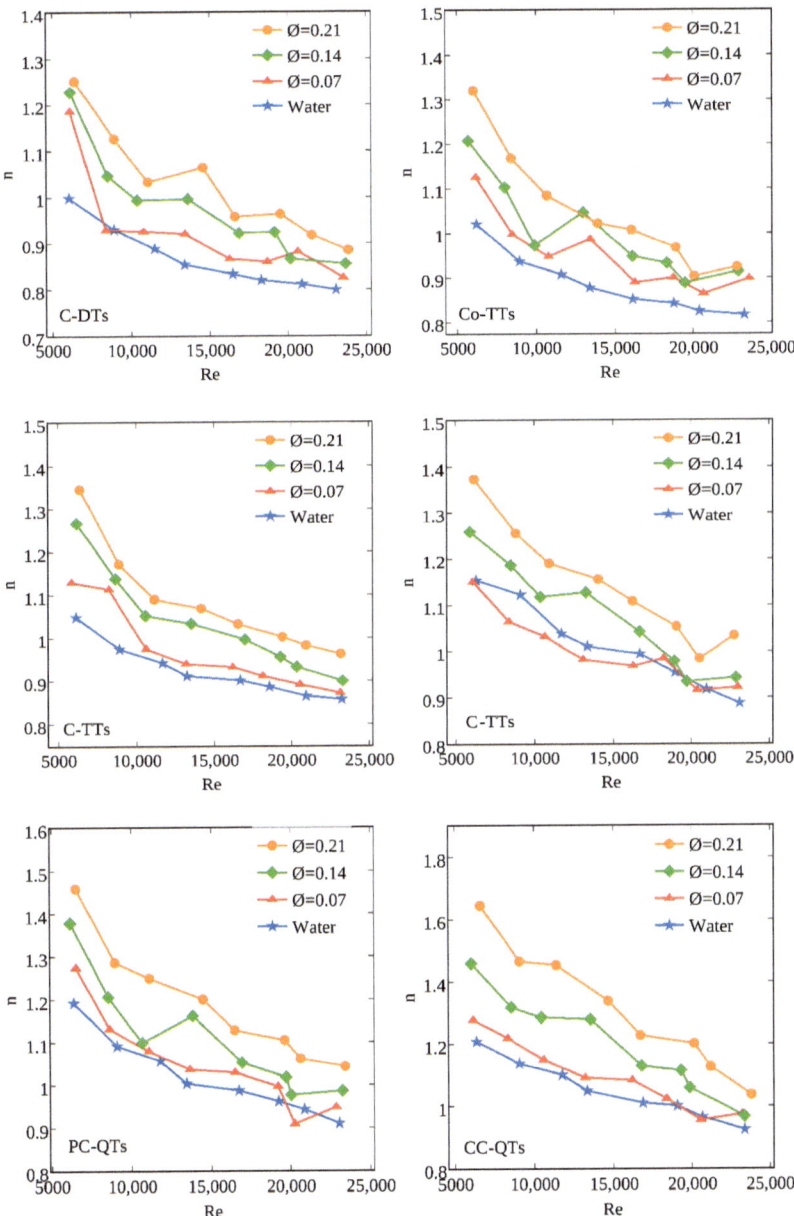

**Fig. 5.12 Reynolds Number variation with Thermal Performance Factor for Nanofluid (cont.)**

This indicates the effect of increasing $Al_2O_3$ nanoparticles on thermal performance factor was more pronounced for the heat transfer improvement as positive effect than friction loss as negative effect in a range of Reynolds number 5,000–25,000, especially at lower Reynolds number where pressure losses were insignificant. As nanofluid with $Al_2O_3$ concentration was used, the thermal performance of Co-QTs, PC-QTs and CC-QTs was 1.03-1.37, 1.04-1.45 and 1.04-1.64, respectively.

# CHAPTER 6
# CONCLUSIONS AND FUTURE SCOPES

This Chapter concludes the presented research study and provides the recommendation for the future study.

## 6.1 CONCLUSIONS

The objective of this work is to perform the comprehensive study on the multiple twisted tape inserts with water and $Al_2O_3$ nanofluid. Performance investigation of multiple twisted tape inserts are carried out at constant heat flux condition with mass flow rate variation of fluid for number of tapes and tape combinations such as ST, Co-DTs, Co-TTs, Co-QTs, C-DTs, C-TTs, PC-QTs and CC-QTs. This work also includes the effect of variation in the volume concentration of nanoparticle in water. For this purpose, eight combinations of multiple twisted tapes and three different concentration of nanofluids were tested for tubular heat exchanger. From the experimental investigations, following conclusions are made.

1. For water and different concentration of nanofluids, Nusselt number, friction factor and thermal performance factor increases with increase in number of tapes.
2. For water and different concentration of nanofluids, the tapes in counter arrangement provides higher thermal performance factor than that in co-arrangement, Interestingly, CC-QTs exhibits superior twisted tape which delivers not only high Nusselt number but also high thermal performance factor.
3. For water, within the range of Reynolds number 5000–25,000, Nusselt number in tubes with the Typical single twisted tape, Co-DTs, C-DTs, Co-TTs, C-TTs, Co-QTs, PC-QTs and CC-QTs were found 13-20%, 18-28%, 31-48%, 50-68%, 63-79%, 80-108%, 86-108% and 100-133% respectively higher than that compared to the plain tube. The enhancement of Nusselt number is also accompanied with friction factors around 1.7-2.6, 3.4-5.3, 3.9-5.5, 5.2-7.9, 5.5-8.1, 6.8-10.4, 7.2-10.8 and 7.4-11.2 times higher than that of the plain tube.
4. For water, thermal performance factors for typical single twisted tape, Co-DTs, C-DTs, Co-TTs, C-TTs, Co-QTs, PC-QTs and CC-QTs, were found to be 0.71-0.94, 0.76-0.96, 0.80-0.1, 0.82-1.02, 0.86-1.05, 0.89-1.15, 0.91-1.19 and 0.92-1.21 respectively.

5. For water and all concentrations of nanofluid, Nusselt number and friction factor increases with increase in mass flow rate whereas, thermal performance factor decreases with increase in mass flow rate.
6. For water and all concentrations of nanofluid, the ratio of Nusselt number of the tube with all combinations of twisted tape inserts to the Nusselt number of the tube without twisted tape insert $\left(N_{ut}/N_{np}\right)$ decreases with increase in mass flow rate.
7. For water and all concentrations of nanofluid, the ratio of friction factor of tube with all combinations of twisted tape inserts to the friction factor of tube without twisted tape insert $\left(f_t/f_p\right)$ increases with increase in mass flow rate.
8. The ratios, $\left(N_{ut}/N_{np}\right)$ and $\left(f_t/f_p\right)$ increases with increase in number of tapes as well as volume concentration of nanoparticle for the same mass flow rate.
9. The Nusselt number of nanofluid with $Al_2O_3$ concentration of 0.07%, 0.14% and 0.21% by volume were respectively 4-8%, 8.4-13.2% and 15.3-24.9% higher than that of the base fluid (Water).
10. The friction factor of nanofluid with $Al_2O_3$ concentration of 0.07%, 0.14% and 0.21% by volume were respectively 2.5-5.9%, 7.2-10.9% and 11.8-17.4% higher than that of the base fluid (Water).
11. The Thermal performance factor of nanofluid with $Al_2O_3$ concentration of 0.07%, 0.14% and 0.21% by volume were respectively 6.7-18.5%, 17-24.7% and 21.6-31.7% higher than that of the base fluid (Water).

## 6.2 FUTURE SCOPE

Investigation of thermal characteristics of $Al_2O_3$/water nanofluid with and without multiple twisted tape inserts in tubular heat exchanger is a challenging task. As this work is the limited study of the few parameters that affects the performance, it can be further extended with the future scope which are given as below.

1. In this work, heat flux is kept constant and the affecting parameters are studied. In the future one can study the effect of changing heat flux on performance of heat exchanger and twisted tapes.

2. This work was limited to the single, dual, triple and quadruple twisted tapes. One can study effect of other possible combinations of twisted tapes.
3. In addition to multiple twisted tapes, effect of different other passive devices like spiral fin, helical springs, snail entry, axial/ radial guide vane can also be studied.
4. The performance investigation of heat exchanger with and without twisted tapes can be carried out in future by replacing $Al_2O_3$ nanofluid, which decides the feasibility of use of alternative nanofluid, effect of different surfactant in base fluid, different orientations of heat exchanger can also be studied.

# REFERENCES

1. M.M.K. Bhuiya, M.S.U. Chowdhury, M. Shahabuddin, M. Saha, L.A. Memone, "Thermal characteristics in a heat exchanger tube fitted with triple twisted tape inserts", *International Communications in Heat and Mass Transfer*, Vol. 48 (2013), 124–132

2. Xiaoyu Zhang, Zhichun Liu, Wei Liu, "Numerical studies on heat transfer and flow characteristics for laminar flow in a tube with multiple regularly spaced twisted tapes", *International Journal of Thermal Sciences*, Vol. 58 (2012), 157-167

3. C. Thianpong, S. Eiamsa-ard, P. Eiamsa-ard, "Turbulent heat transfer enhancement by counter/co-swirling flow in a tube fitted with twin twisted tapes", *Experimental Thermal and Fluid Science*, Vol. 34 (2010), 53–62

4. S. Eiamsa-ard, K. Wongcharee, S. Sripattanapipat,"3-D Numerical simulation of swirling flow and convective heat transfer in a circular tube induced by means of loose-fit twisted tapes" *International Communications in Heat and Mass Transfer*, Vol. 36 (2009), 947–955

5. M.M.K. Bhuiya, A.S.M. Sayem, M. Islam, M.S.U. Chowdhury, M. Shahabuddin, "Performance assessment in a heat exchanger tube fitted with double counter twisted tape inserts" *International Communications in Heat and Mass Transfer*, Vol. 50 (2014) 25–33

6. Halit Bas, Veysel Ozceyhan, "Heat transfer enhancement in a tube with twisted tape inserts placed separately from the tube wall", *Experimental Thermal and Fluid Science*, Vol. 41 (2012) 51–58

7. W.H. Azmi, K.V. Sharma, P.K. Sarma, Rizalman Mamat, ShahraniAnuar, "Comparison of convective heat transfer coefficient and friction factor of $TiO_2$ nanofluid flow in a tube with twisted tape inserts", *International Journal of Thermal Sciences*, Vol. 81 (2014), 84-93

8. E. Esmaeilzadeh, H. Almohammadi, A. Nokhosteen, A. Motezaker, A.N. Omrani, "Study on heat transfer and friction factor characteristics of $\Upsilon$-$Al_2O_3$/water through circular tube with twisted tapeinserts with different thicknesses", *International Journal of Thermal Sciences*, Vol. 82 (2014), 72- 83

9. Smith Eiamsa-ard, Kunlanan Kiatkittipong, "Heat transfer enhancement by multiple twisted tape insertsand $TiO_2$/water nanofluid", *Applied Thermal Engineering*, Vol.70 (2014), 896-924

10. M.T. Naik, G. RangaJanardana, L. SyamSundar, "Experimental investigation of heat transfer and friction factor with water–propylene glycol based CuO nanofluid in a tube with twisted tape inserts", *International Communications in Heat and Mass Transfer*, Vol. 46 (2013), 13–21

11. S. Eiamsa-ard, K. Wongcharee, "Single-phase heat transfer of CuO/water nanofluids in micro-fin tube equipped with dual twisted-tapes", *International Communications in Heat and Mass Transfer*, Vol. 39 (2012), 1453–1459

12. L. SyamSundar, N.T. Ravi Kumar, M.T. Naik, K.V. Sharma, "Effect of full length twisted tape inserts on heat transfer and friction factor enhancement with $Fe_3O_4$ magnetic nanofluid inside a plain tube: An experimental study", *International Journal of Heat and Mass Transfer*, Vol. 55 (2012), 2761–2768

13. L. Syam Sundar, K.V. Sharma, "Turbulent heat transfer and friction factor of $Al_2O_3$ Nanofluid in circular tube with twisted tape inserts", *International Journal of Heat and Mass Transfer*, Vol. 53 (2010) 1409–1416

14. Y. Raja Sekhara, K.V.Sharmab, R.Thundil Karupparaja, C. Chiranjeevia, "Heat Transfer Enhancement with $Al_2O_3$ Nanofluids and Twisted Tapes in a Pipe for Solar Thermal Applications", *International Conference on Design and Manufacturing*, 64 (2013) 1474 – 1484

15. W.H. Azmi, K.V. Sharma, P.K. Sarma, Rizalman Mamat, Shahrani Anuar, "Comparison of convective heat transfer coefficient and friction factor of $TiO_2$ nanofluid flow in a tube with twisted tape inserts", *International Journal of Thermal Sciences*, vol. 81 (2014) 84-93

16. E. Esmaeilzadeh*, H. Almohammadi, A. Nokhosteen, A. Motezaker, A.N. Omrani, "Study on heat transfer and friction factor characteristics of $\Upsilon$ -$Al_2O_3$/water through circular tube with twisted tape inserts with different thicknesses", *International Journal of Thermal Sciences*, Vol. 82 (2014) 72-83

17. A.Bejan, A.D. Kraus, *Heat Transfer Handbook*, John Wiley, New Jersey, 2003.

18. F. Giampietro, "Heat transfer optimization in internally finned tubes under laminar flow conditions", *International Journal of Heat Mass Transfer*, Vol. 41 (1998) 1243-1253.

19. P.G. Vicente, A. Garcia, A. Viedma, "Experimental investigation on heat transfer and frictional characteristics of spirally corrugated tubes in turbulent flow at different Prandtl numbers", *International Journal of Heat Mass Transfer*, Vol. 47 (4) (2004) 671-681.

20. P. Naphon, M. Nuchjapo, J. Kurujareon, "Tube side heat transfer coefficient and friction factor characteristics of horizontal tubes with helical rib", *Energy Convers. Manage.* Vol. 47 (2006) 3031-3044.
21. M. Siddique, M. Alhazmy, "Experimental study of turbulent single-phase flow and heat transfer inside a micro-finned tube", *International Journal of Refrigeration*, Vol. 31 (2008) 234-241.
22. J.M. Whitham, "The effects of retarders in fire tubes of steam boilers", *Str. Railw.J.* Vol. 12 (1896) 374.
23. R.L. Webb, "Performance evaluation criteria for use of enhanced heat transfer surfaces in heat exchanger design", *International Journal of Heat Mass Transfer.* Vol. 24 (1981) 715-726.
24. G.C. Kidd Jr., "Heat transfer and pressure drop for nitrogen flowing in tubes containing twisted tapes", *AIChE J.* Vol. 15 (1969) 581-585.
25. L. Wang, B. Sunden, "Performance comparison of some tube inserts", *International Communication in Heat Mass Transf.* Vol. 29 (2002) 45-56.
26. O.H. Klepper, Heat transfer performance of short twisted tapes, *AIChE J.* Vol. 35 (1972) 1-24.
27. R.M. Manglik, A.E. Bergles, "Heat transfer and pressure drop correlations for twisted-tape inserts in isothermal tubes". *ASME Journal of Heat Transfer*, Vol. 115 (1993) 890-896.
28. R.L. Webb, "Performance evaluation criteria for use of enhanced heat transfer surfaces in heat exchanger design", *International Journal of Heat Mass Transfer.* Vol. 24 (1981) 715-726.
29. G.E. Kondhalkar, V.N. kapatkat, "Performance Analysis of Spiral Tube Heat Exchanger used in Oil extraction system", *International Journal of Modern Engineering Research*, Vol.2, Issue 3, May-June(2012) 930-936
30. H. Almohammadi, S.N. Vatan, "Experimental Investigation of Convective Heat Transfer and Pressure Drop of $Al_2O_3$/Water Nanofluid in Laminar Flow Regime inside a Circular Tube", *World Academy of Science, Engineering and Technology*, Vol.68 (2012)
31. F.M.Pakdaman, A.A.M.Bahabadi, " An Experimental Investigation on Thermo-physical Properties and Overall Performance of MWCNT/Heat Transfer Oil Nanofluid Inside Vertical Helically Coiled Tubes ", *Experimental Thermal and Fluid Science*, Vol.40 (2012), 103-111

32. Yunuar, N. Putra, Gunawan M. Baqi, "Flow and convective heat transfer characteristics of spiral pipe for nanofluid", IJRRAS Vol.7, June 2011
33. F.M. Pakdaman, A.A.M.Behabadi, P.Razi, "An experimental investigation on thermo-physical properties and overall performance of MWCNT/heat transfer oil nanofluid flow inside vertical helically coiled tubes", *Experimental thermal and fluid science*, Vol.40 (2012) 103-111

# APPENDIX A

# Properties of Water

### Table A-1 Water Properties at Different Temperatures

| Temp (K) | $\mu$ $10^{-3} \left(\frac{Ns}{m^2}\right)$ | $\varrho$ $\left(\frac{kg}{m^3}\right)$ | Cp $\left(\frac{J}{kg\,K}\right)$ | k $\left(\frac{W}{mK}\right)$ |
|---|---|---|---|---|
| 298 | 0.891 | 997.0 | 4180.0 | 0.607 |
| 299 | 0.873 | 997.0 | 4179.5 | 0.610 |
| 300 | 0.855 | 997.0 | 4179.0 | 0.613 |
| 301 | 0.836 | 996.7 | 4178.7 | 0.614 |
| 302 | 0.817 | 996.3 | 4178.3 | 0.614 |
| 303 | 0.798 | 996.0 | 4178.0 | 0.615 |
| 304 | 0.779 | 995.5 | 4178.0 | 0.618 |
| 305 | 0.760 | 995.0 | 4178.0 | 0.620 |
| 306 | 0.747 | 994.7 | 4178.0 | 0.621 |
| 307 | 0.733 | 994.3 | 4178.0 | 0.622 |
| 308 | 0.720 | 994.0 | 4178.0 | 0.623 |
| 309 | 0.708 | 993.5 | 4178.0 | 0.626 |
| 310 | 0.695 | 993.0 | 4178.0 | 0.628 |
| 311 | 0.681 | 992.7 | 4178.3 | 0.629 |
| 312 | 0.667 | 992.4 | 4178.7 | 0.630 |
| 313 | 0.653 | 992.1 | 4179.0 | 0.631 |
| 314 | 0.642 | 991.6 | 4179.0 | 0.633 |
| 315 | 0.631 | 991.1 | 4179.0 | 0.634 |
| 316 | 0.619 | 990.8 | 4179.3 | 0.635 |
| 317 | 0.608 | 990.4 | 4179.7 | 0.636 |
| 318 | 0.596 | 990.1 | 4180.0 | 0.637 |
| 319 | 0.587 | 989.6 | 4180.0 | 0.639 |
| 320 | 0.577 | 989.1 | 4180.0 | 0.640 |
| 321 | 0.567 | 988.8 | 4180.3 | 0.641 |
| 322 | 0.557 | 988.4 | 4180.7 | 0.643 |
| 323 | 0.547 | 988.1 | 4181.0 | 0.644 |
| 324 | 0.538 | 987.7 | 4181.5 | 0.645 |
| 325 | 0.528 | 987.2 | 4182.0 | 0.645 |
| 326 | 0.520 | 986.5 | 4182.3 | 0.646 |
| 327 | 0.512 | 985.9 | 4182.7 | 0.648 |
| 328 | 0.504 | 985.2 | 4183.0 | 0.649 |
| 329 | 0.497 | 984.8 | 4183.5 | 0.650 |
| 330 | 0.489 | 984.3 | 4184.0 | 0.650 |
| 331 | 0.482 | 984.0 | 4184.3 | 0.651 |
| 332 | 0.474 | 983.6 | 4184.7 | 0.653 |
| 333 | 0.467 | 983.3 | 4185.0 | 0.654 |

| | | | | |
|---|---|---|---|---|
| 334 | 0.460 | 982.8 | 4185.5 | 0.655 |
| 335 | 0.453 | 982.3 | 4186.0 | 0.656 |
| 336 | 0.446 | 981.7 | 4186.3 | 0.657 |
| 337 | 0.440 | 981.0 | 4186.7 | 0.658 |
| 338 | 0.433 | 980.4 | 4187.0 | 0.659 |
| 339 | 0.427 | 979.9 | 4187.5 | 0.660 |
| 340 | 0.420 | 979.4 | 4188.0 | 0.660 |
| 341 | 0.415 | 978.8 | 4188.7 | 0.661 |
| 342 | 0.409 | 978.1 | 4189.3 | 0.662 |
| 343 | 0.404 | 977.5 | 4190.0 | 0.663 |
| 344 | 0.399 | 977.0 | 4190.2 | 0.664 |
| 345 | 0.394 | 976.5 | 4190.4 | 0.665 |
| 346 | 0.388 | 976.0 | 4190.6 | 0.666 |
| 347 | 0.383 | 975.5 | 4190.8 | 0.667 |
| 348 | 0.378 | 975.0 | 4191.0 | 0.668 |
| 349 | 0.373 | 974.4 | 4192.4 | 0.668 |
| 350 | 0.369 | 973.8 | 4193.8 | 0.668 |
| 351 | 0.364 | 973.2 | 4195.2 | 0.669 |
| 352 | 0.360 | 972.6 | 4196.6 | 0.669 |
| 353 | 0.355 | 972.0 | 4198.0 | 0.670 |
| 354 | 0.351 | 971.2 | 4199.0 | 0.670 |
| 355 | 0.347 | 970.4 | 4200.0 | 0.671 |
| 356 | 0.342 | 969.6 | 4201.0 | 0.672 |
| 357 | 0.338 | 968.8 | 4202.0 | 0.672 |
| 358 | 0.334 | 968.0 | 4203.0 | 0.673 |
| 359 | 0.330 | 966.8 | 4204.0 | 0.673 |
| 360 | 0.326 | 965.6 | 4205.0 | 0.674 |
| 361 | 0.322 | 964.4 | 4206.0 | 0.675 |
| 362 | 0.318 | 963.2 | 4207.0 | 0.675 |

**Properties of Al$_2$O$_3$ Nanofluid is Calculated as:**

**A) For 0.07% Volume Concentration of Al$_2$O$_3$ in Distilled Water at 30$^0$C**

1) Density of Nanofluid:

$$\rho_{nf} = \emptyset \rho_p + (1 - \emptyset)\rho_w$$

$$\rho_{nf} = (0.07)(3950) + (1 - 0.07)(996)$$

$$\rho_{nf} = 1202.78 \; kg/m^3$$

2) Specific Heat of Nanofluid:

$$C_{nf} = \frac{\emptyset(\rho\,C)_p + (1 - \emptyset)(\rho\,C)_w}{\emptyset \rho_p + (1 - \emptyset)\rho_w}$$

$$C_{nf} = \frac{(0.07)(3950)(784.9) + (1 - 0.07)(996)(4178)}{(0.07)(3950) + (1 - 0.07)(996)}$$

$$C_{nf} = 3397.98 \; \text{J/kg K}$$

3) Thermal Conductivity of Nanofluid:

$$\frac{k_{nf}}{k_w} = \frac{k_p + 2k_w - 2\emptyset\,(k_w - k_p)}{k_p + 2k_w + \emptyset\,(k_w - k_p)}$$

$$\frac{k_{nf}}{0.615} = \frac{(40) + (2)(0.615) - (2)(0.07)(0.615 - 40)}{(40) + (2)(0.615) + (0.07)(0.615 - 40)}$$

$$k_{nf} = 0.747 \; W/mK$$

4) Dynamic Viscosity of Nanofluid:

$$\frac{\mu_{nf}}{\mu_w} = 1 + 2.5\,\emptyset$$

$$\frac{\mu_{nf}}{0.798\,e^{-3}} = 1 + (2.5)(0.07)$$

$$\mu_{nf} = 0.938 e^{-3} \; N - s/m^2$$

## B) For 0.14% Volume concentration of Al$_2$O$_3$ in Distilled Water at 30°C

1) Density of Nanofluid:

$$\rho_{nf} = \emptyset\rho_p + (1 - \emptyset)\rho_w$$

$$\rho_{nf} = (0.14)(3950) + (1 - 0.14)(996)$$

$$\rho_{nf} = 1409.56 \; kg/m^3$$

2) Specific Heat of Nanofluid:

$$C_{nf} = \frac{\emptyset(\rho \, C)_p + (1 - \emptyset)(\rho \, C)_w}{\emptyset\rho_p + (1 - \emptyset)\rho_w}$$

$$C_{nf} = \frac{(0.14)(3950)(784.9) + (1 - 0.14)(996)(4178)}{(0.14)(3950) + (1 - 0.14)(996)}$$

$$C_{nf} = 2846.8 \; J/kg \; K$$

3) Thermal Conductivity of Nanofluid:

$$\frac{k_{nf}}{k_w} = \frac{k_p + 2k_w - 2\emptyset(k_w - k_p)}{k_p + 2k_w + \emptyset(k_w - k_p)}$$

$$\frac{k_{nf}}{0.615} = \frac{(40) + (2)(0.615) - (2)(0.14)(0.615 - 40)}{(40) + (2)(0.615) + (0.14)(0.615 - 40)}$$

$$k_{nf} = 0.9 \; W/mK$$

4) Dynamic Viscosity of Nanofluid:

$$\frac{\mu_{nf}}{\mu_w} = 1 + 2.5 \, \emptyset$$

$$\frac{\mu_{nf}}{0.798 \, e^{-3}} = 1 + (2.5)(0.14)$$

$$\mu_{nf} = 1.077 e^{-3} \; N - s/m^2$$

## C) For 0.21% Volume Concentration of Al$_2$O$_3$ in Distilled Water at 30°C

1) Density of Nanofluid:

$$\rho_{nf} = \emptyset \rho_p + (1 - \emptyset)\rho_w$$

$$\rho_{nf} = (0.21)(3950) + (1 - 0.21)(996)$$

$$\rho_{nf} = 1616.34 \, kg/m^3$$

2) Specific Heat of Nanofluid:

$$C_{nf} = \frac{\emptyset(\rho \, C)_p + (1 - \emptyset)(\rho \, C)_w}{\emptyset \rho_p + (1 - \emptyset)\rho_w}$$

$$C_{nf} = \frac{(0.21)(3950)(784.9) + (1 - 0.21)(996)(4178)}{(0.21)(3950) + (1 - 0.21)(996)}$$

$$C_{nf} = 2436.68 \, J/kg \, K$$

3) Thermal Conductivity of Nanofluid:

$$\frac{k_{nf}}{k_w} = \frac{k_p + 2k_w - 2\emptyset \, (k_w - k_p)}{k_p + 2k_w + \emptyset \, (k_w - k_p)}$$

$$\frac{k_{nf}}{0.615} = \frac{(40) + (2)(0.615) - (2)(0.21)(0.615 - 40)}{(40) + (2)(0.615) + (0.21)(0.615 - 40)}$$

$$k_{nf} = 1.078 \, W/mK$$

4) Dynamic Viscosity of Nanofluid:

$$\frac{\mu_{nf}}{\mu_w} = 1 + 2.5 \, \emptyset$$

$$\frac{\mu_{nf}}{0.798 \, e^{-3}} = 1 + (2.5)(0.21)$$

$$\mu_{nf} = 1.217 e^{-3} \, N - s/m^2$$

## Properties of Al$_2$O$_3$ Nanofluid

### Table A-2 Properties of Al$_2$O$_3$ Nanofluid for 0.07% Volume Concentration

| Temp (K) | $\mu$ $10^{-3}\left(\frac{Ns}{m^2}\right)$ | $\rho$ $\left(\frac{kg}{m^3}\right)$ | Cp $\left(\frac{J}{kg\,K}\right)$ | k $\left(\frac{W}{mK}\right)$ |
|---|---|---|---|---|
| 298 | 1.047 | 1203.7 | 3397.6 | 0.738 |
| 299 | 1.026 | 1203.7 | 3397.7 | 0.741 |
| 300 | 1.005 | 1203.7 | 3397.9 | 0.745 |
| 301 | 0.982 | 1203.4 | 3398.0 | 0.746 |
| 302 | 0.960 | 1203.1 | 3398.0 | 0.746 |
| 303 | 0.938 | 1202.8 | 3398.0 | 0.747 |
| 304 | 0.915 | 1202.3 | 3398.2 | 0.750 |
| 305 | 0.893 | 1201.9 | 3398.3 | 0.753 |
| 306 | 0.877 | 1201.5 | 3398.6 | 0.754 |
| 307 | 0.862 | 1201.2 | 3398.9 | 0.756 |
| 308 | 0.846 | 1200.9 | 3399.1 | 0.757 |
| 309 | 0.831 | 1200.5 | 3399.3 | 0.760 |
| 310 | 0.817 | 1200.0 | 3399.5 | 0.763 |
| 311 | 0.800 | 1199.7 | 3400.0 | 0.764 |
| 312 | 0.784 | 1199.4 | 3400.5 | 0.765 |
| 313 | 0.767 | 1199.2 | 3401.0 | 0.766 |
| 314 | 0.754 | 1198.7 | 3401.1 | 0.768 |
| 315 | 0.741 | 1198.2 | 3401.3 | 0.770 |
| 316 | 0.728 | 1197.9 | 3401.7 | 0.771 |
| 317 | 0.714 | 1197.6 | 3402.2 | 0.772 |
| 318 | 0.700 | 1197.3 | 3402.7 | 0.774 |
| 319 | 0.689 | 1196.8 | 3402.8 | 0.776 |
| 320 | 0.678 | 1196.4 | 3403.0 | 0.777 |
| 321 | 0.666 | 1196.1 | 3403.5 | 0.779 |
| 322 | 0.654 | 1195.7 | 3403.9 | 0.781 |
| 323 | 0.643 | 1195.4 | 3404.4 | 0.782 |
| 324 | 0.632 | 1195.0 | 3405.0 | 0.783 |
| 325 | 0.620 | 1194.6 | 3405.5 | 0.783 |
| 326 | 0.611 | 1194.0 | 3405.8 | 0.785 |
| 327 | 0.602 | 1193.4 | 3406.1 | 0.787 |
| 328 | 0.592 | 1192.7 | 3406.4 | 0.788 |
| 329 | 0.583 | 1192.3 | 3406.9 | 0.789 |
| 330 | 0.575 | 1191.9 | 3407.5 | 0.789 |
| 331 | 0.566 | 1191.6 | 3407.9 | 0.791 |
| 332 | 0.557 | 1191.3 | 3408.4 | 0.793 |
| 333 | 0.549 | 1191.0 | 3408.9 | 0.794 |

| | | | | |
|---|---|---|---|---|
| 334 | 0.541 | 1190.5 | 3409.3 | 0.795 |
| 335 | 0.532 | 1190.0 | 3409.8 | 0.797 |
| 336 | 0.524 | 1189.5 | 3410.1 | 0.798 |
| 337 | 0.517 | 1188.9 | 3410.4 | 0.799 |
| 338 | 0.509 | 1188.3 | 3410.7 | 0.800 |
| 339 | 0.501 | 1187.8 | 3411.2 | 0.801 |
| 340 | 0.494 | 1187.3 | 3411.7 | 0.801 |
| 341 | 0.487 | 1186.8 | 3412.2 | 0.803 |
| 342 | 0.481 | 1186.2 | 3412.7 | 0.804 |
| 343 | 0.475 | 1185.6 | 3413.3 | 0.805 |
| 344 | 0.469 | 1185.1 | 3413.5 | 0.806 |
| 345 | 0.462 | 1184.6 | 3413.8 | 0.807 |
| 346 | 0.456 | 1184.2 | 3414.1 | 0.809 |
| 347 | 0.450 | 1183.7 | 3414.3 | 0.810 |
| 348 | 0.444 | 1183.3 | 3414.6 | 0.811 |
| 349 | 0.439 | 1182.7 | 3415.7 | 0.811 |
| 350 | 0.433 | 1182.1 | 3416.8 | 0.811 |
| 351 | 0.428 | 1181.6 | 3417.9 | 0.812 |
| 352 | 0.423 | 1181.0 | 3419.0 | 0.812 |
| 353 | 0.417 | 1180.5 | 3420.2 | 0.813 |
| 354 | 0.412 | 1179.7 | 3420.9 | 0.814 |
| 355 | 0.407 | 1179.0 | 3421.5 | 0.815 |
| 356 | 0.402 | 1178.2 | 3422.2 | 0.815 |
| 357 | 0.397 | 1177.5 | 3422.9 | 0.816 |
| 358 | 0.392 | 1176.7 | 3423.6 | 0.817 |
| 359 | 0.388 | 1175.6 | 3424.0 | 0.817 |
| 360 | 0.383 | 1174.5 | 3424.5 | 0.818 |
| 361 | 0.378 | 1173.4 | 3424.9 | 0.819 |
| 362 | 0.374 | 1172.3 | 3425.4 | 0.820 |
| 363 | 0.369 | 1171.2 | 3425.8 | 0.820 |

## Table A-3 Properties of Al$_2$O$_3$ Nanofluid for 0.14% Volume Concentration

| Temp (K) | $\mu$ $10^{-3}\left(\frac{Ns}{m^2}\right)$ | $\varrho$ $\left(\frac{kg}{m^3}\right)$ | Cp $\left(\frac{J}{kg\,K}\right)$ | k $\left(\frac{W}{mK}\right)$ |
|---|---|---|---|---|
| 298 | 1.203 | 1410.4 | 2844.6 | 0.888 |
| 299 | 1.179 | 1410.4 | 2845.1 | 0.893 |
| 300 | 1.154 | 1410.4 | 2845.8 | 0.897 |
| 301 | 1.129 | 1410.1 | 2846.2 | 0.898 |
| 302 | 1.103 | 1409.8 | 2846.5 | 0.899 |
| 303 | 1.077 | 1409.6 | 2846.8 | 0.900 |
| 304 | 1.052 | 1409.1 | 2847.2 | 0.903 |
| 305 | 1.026 | 1408.7 | 2847.6 | 0.907 |
| 306 | 1.008 | 1408.4 | 2848.2 | 0.908 |
| 307 | 0.990 | 1408.1 | 2848.7 | 0.910 |
| 308 | 0.972 | 1407.8 | 2849.2 | 0.911 |
| 309 | 0.955 | 1407.4 | 2849.6 | 0.915 |
| 310 | 0.938 | 1407.0 | 2850.0 | 0.919 |
| 311 | 0.919 | 1406.7 | 2850.7 | 0.920 |
| 312 | 0.900 | 1406.5 | 2851.4 | 0.921 |
| 313 | 0.882 | 1406.2 | 2852.1 | 0.923 |
| 314 | 0.867 | 1405.8 | 2852.4 | 0.925 |
| 315 | 0.852 | 1405.3 | 2852.8 | 0.927 |
| 316 | 0.836 | 1405.1 | 2853.4 | 0.929 |
| 317 | 0.820 | 1404.8 | 2854.1 | 0.930 |
| 318 | 0.805 | 1404.5 | 2854.8 | 0.931 |
| 319 | 0.792 | 1404.1 | 2855.1 | 0.934 |
| 320 | 0.779 | 1403.6 | 2855.4 | 0.936 |
| 321 | 0.765 | 1403.3 | 2856.1 | 0.938 |
| 322 | 0.752 | 1403.1 | 2856.8 | 0.940 |
| 323 | 0.738 | 1402.8 | 2857.4 | 0.942 |
| 324 | 0.726 | 1402.4 | 2858.1 | 0.942 |
| 325 | 0.713 | 1402.0 | 2858.8 | 0.943 |
| 326 | 0.702 | 1401.4 | 2859.2 | 0.945 |
| 327 | 0.691 | 1400.8 | 2859.6 | 0.947 |
| 328 | 0.680 | 1400.3 | 2860.0 | 0.949 |
| 329 | 0.670 | 1399.9 | 2860.6 | 0.949 |
| 330 | 0.660 | 1399.5 | 2861.3 | 0.950 |
| 331 | 0.650 | 1399.2 | 2861.9 | 0.952 |
| 332 | 0.640 | 1398.9 | 2862.6 | 0.954 |
| 333 | 0.630 | 1398.6 | 2863.2 | 0.956 |
| 334 | 0.621 | 1398.2 | 2863.8 | 0.957 |
| 335 | 0.612 | 1397.8 | 2864.4 | 0.959 |

| | | | | |
|---|---|---|---|---|
| 336 | 0.603 | 1397.2 | 2864.8 | 0.960 |
| 337 | 0.594 | 1396.7 | 2865.1 | 0.962 |
| 338 | 0.585 | 1396.1 | 2865.5 | 0.963 |
| 339 | 0.576 | 1395.7 | 2866.1 | 0.964 |
| 340 | 0.567 | 1395.3 | 2866.7 | 0.965 |
| 341 | 0.560 | 1394.7 | 2867.3 | 0.966 |
| 342 | 0.553 | 1394.2 | 2867.9 | 0.967 |
| 343 | 0.545 | 1393.7 | 2868.5 | 0.969 |
| 344 | 0.538 | 1393.2 | 2868.9 | 0.970 |
| 345 | 0.531 | 1392.8 | 2869.3 | 0.972 |
| 346 | 0.524 | 1392.4 | 2869.7 | 0.973 |
| 347 | 0.517 | 1391.9 | 2870.1 | 0.975 |
| 348 | 0.510 | 1391.5 | 2870.5 | 0.976 |
| 349 | 0.504 | 1391.0 | 2871.6 | 0.976 |
| 350 | 0.498 | 1390.5 | 2872.7 | 0.976 |
| 351 | 0.492 | 1390.0 | 2873.7 | 0.977 |
| 352 | 0.485 | 1389.4 | 2874.8 | 0.978 |
| 353 | 0.479 | 1388.9 | 2875.8 | 0.979 |
| 354 | 0.474 | 1388.2 | 2876.5 | 0.979 |
| 355 | 0.468 | 1387.5 | 2877.1 | 0.980 |
| 356 | 0.462 | 1386.9 | 2877.8 | 0.981 |
| 357 | 0.457 | 1386.2 | 2878.4 | 0.982 |
| 358 | 0.451 | 1385.5 | 2879.1 | 0.983 |
| 359 | 0.446 | 1384.4 | 2879.4 | 0.984 |
| 360 | 0.440 | 1383.4 | 2879.7 | 0.985 |
| 361 | 0.435 | 1382.4 | 2880.0 | 0.985 |
| 362 | 0.429 | 1381.4 | 2880.3 | 0.986 |
| 363 | 0.424 | 1380.3 | 2880.7 | 0.987 |

## Table A-4 Properties of Al$_2$O$_3$ Nanofluid for 0.21% Volume Concentration

| Temp (K) | μ $10^{-3}\left(\frac{Ns}{m^2}\right)$ | ρ $\left(\frac{kg}{m^3}\right)$ | Cp $\left(\frac{J}{kg\,K}\right)$ | k $\left(\frac{W}{mK}\right)$ |
|---|---|---|---|---|
| 298 | 1.359 | 1617.130 | 2432.970 | 1.064 |
| 299 | 1.331 | 1617.130 | 2433.802 | 1.069 |
| 300 | 1.304 | 1617.130 | 2434.849 | 1.075 |
| 301 | 1.275 | 1616.867 | 2435.460 | 1.076 |
| 302 | 1.246 | 1616.603 | 2436.070 | 1.077 |
| 303 | 1.217 | 1616.340 | 2436.681 | 1.078 |
| 304 | 1.188 | 1615.945 | 2437.313 | 1.082 |
| 305 | 1.159 | 1615.550 | 2437.945 | 1.087 |
| 306 | 1.139 | 1615.287 | 2438.719 | 1.088 |
| 307 | 1.118 | 1615.023 | 2439.493 | 1.090 |
| 308 | 1.098 | 1614.760 | 2440.267 | 1.092 |
| 309 | 1.079 | 1614.365 | 2440.900 | 1.096 |
| 310 | 1.060 | 1613.970 | 2441.534 | 1.100 |
| 311 | 1.039 | 1613.733 | 2442.399 | 1.102 |
| 312 | 1.017 | 1613.496 | 2443.264 | 1.104 |
| 313 | 0.996 | 1613.259 | 2444.129 | 1.105 |
| 314 | 0.979 | 1612.864 | 2444.662 | 1.108 |
| 315 | 0.962 | 1612.469 | 2445.196 | 1.110 |
| 316 | 0.944 | 1612.206 | 2446.034 | 1.112 |
| 317 | 0.927 | 1611.942 | 2446.871 | 1.114 |
| 318 | 0.909 | 1611.679 | 2447.709 | 1.116 |
| 319 | 0.894 | 1611.284 | 2448.244 | 1.118 |
| 320 | 0.880 | 1610.889 | 2448.778 | 1.121 |
| 321 | 0.865 | 1610.626 | 2449.617 | 1.123 |
| 322 | 0.849 | 1610.362 | 2450.455 | 1.125 |
| 323 | 0.834 | 1610.099 | 2451.294 | 1.128 |
| 324 | 0.820 | 1609.744 | 2452.115 | 1.128 |
| 325 | 0.805 | 1609.388 | 2452.935 | 1.129 |
| 326 | 0.793 | 1608.861 | 2453.492 | 1.132 |
| 327 | 0.781 | 1608.335 | 2454.048 | 1.134 |
| 328 | 0.769 | 1607.808 | 2454.604 | 1.136 |
| 329 | 0.757 | 1607.453 | 2455.426 | 1.137 |
| 330 | 0.746 | 1607.097 | 2456.247 | 1.138 |
| 331 | 0.735 | 1606.834 | 2457.037 | 1.140 |
| 332 | 0.723 | 1606.570 | 2457.826 | 1.142 |
| 333 | 0.712 | 1606.307 | 2458.616 | 1.145 |
| 334 | 0.702 | 1605.912 | 2459.345 | 1.146 |
| 335 | 0.691 | 1605.517 | 2460.074 | 1.148 |

| | | | | |
|---|---|---|---|---|
| 336 | 0.681 | 1605.017 | 2460.610 | 1.150 |
| 337 | 0.670 | 1604.516 | 2461.145 | 1.151 |
| 338 | 0.660 | 1604.016 | 2461.681 | 1.153 |
| 339 | 0.650 | 1603.621 | 2462.410 | 1.154 |
| 340 | 0.641 | 1603.226 | 2463.140 | 1.155 |
| 341 | 0.632 | 1602.726 | 2463.837 | 1.157 |
| 342 | 0.624 | 1602.225 | 2464.533 | 1.158 |
| 343 | 0.616 | 1601.725 | 2465.230 | 1.160 |
| 344 | 0.608 | 1601.330 | 2465.815 | 1.162 |
| 345 | 0.600 | 1600.935 | 2466.401 | 1.163 |
| 346 | 0.592 | 1600.540 | 2466.986 | 1.165 |
| 347 | 0.584 | 1600.145 | 2467.571 | 1.167 |
| 348 | 0.576 | 1599.750 | 2468.157 | 1.168 |
| 349 | 0.569 | 1599.276 | 2469.235 | 1.168 |
| 350 | 0.562 | 1598.802 | 2470.313 | 1.168 |
| 351 | 0.555 | 1598.328 | 2471.390 | 1.169 |
| 352 | 0.548 | 1597.854 | 2472.467 | 1.170 |
| 353 | 0.541 | 1597.380 | 2473.544 | 1.171 |
| 354 | 0.535 | 1596.748 | 2474.258 | 1.173 |
| 355 | 0.529 | 1596.116 | 2474.972 | 1.174 |
| 356 | 0.522 | 1595.484 | 2475.684 | 1.175 |
| 357 | 0.516 | 1594.852 | 2476.396 | 1.176 |
| 358 | 0.509 | 1594.220 | 2477.109 | 1.177 |
| 359 | 0.503 | 1593.272 | 2477.478 | 1.178 |
| 360 | 0.497 | 1592.324 | 2477.847 | 1.179 |
| 361 | 0.491 | 1591.376 | 2478.224 | 1.180 |
| 362 | 0.485 | 1590.428 | 2478.600 | 1.181 |
| 363 | 0.479 | 1589.480 | 2478.976 | 1.182 |

# ANNEXURE I

## Measured Parameter of Plain Tube

### For Water

| Sr. No. | $t_1$ (Sec) | $t_2$ (Sec) | $t_3$ (Sec) | $t_{avg}$ (Sec) | $T_{in}$ (°C) | $T_{out}$ (°C) | $T_1$ (°C) | $T_2$ (°C) | $T_3$ (°C) | $T_s$ (°C) | $\Delta P$ (Pa) |
|---|---|---|---|---|---|---|---|---|---|---|---|
| 1 | 24.08 | 24.00 | 23.92 | 24.00 | 55.0 | 67.0 | 90 | 91 | 103 | 94.55 | 24.0 |
| 2 | 17.89 | 17.63 | 17.94 | 17.82 | 60.5 | 70.8 | 87 | 90 | 103 | 93.20 | 38.1 |
| 3 | 14.70 | 14.91 | 14.88 | 14.83 | 63.3 | 72.4 | 87 | 88 | 102 | 92.19 | 51.8 |
| 4 | 12.18 | 11.84 | 11.99 | 12.00 | 65.6 | 73.1 | 84 | 86 | 101 | 90.19 | 72.1 |
| 5 | 9.99 | 10.02 | 9.60 | 9.87 | 68.0 | 74.3 | 82 | 85 | 101 | 89.19 | 101.0 |
| 6 | 8.88 | 8.66 | 8.72 | 8.75 | 70.3 | 75.6 | 81 | 84 | 100 | 88.19 | 119.0 |
| 7 | 7.80 | 7.75 | 7.57 | 7.71 | 72.7 | 77.0 | 80 | 83 | 100 | 87.54 | 146.0 |
| 8 | 6.99 | 7.12 | 6.66 | 6.92 | 75.0 | 78.3 | 79 | 82 | 99 | 86.56 | 168.6 |

### For 0.07% of $Al_2O_3$ Nanofluid

| Sr. No. | $t_1$ (Sec) | $t_2$ (Sec) | $t_3$ (Sec) | $t_{avg}$ (Sec) | $T_{in}$ (°C) | $T_{out}$ (°C) | $T_1$ (°C) | $T_2$ (°C) | $T_3$ (°C) | $T_s$ (°C) | $\Delta P$ (Pa) |
|---|---|---|---|---|---|---|---|---|---|---|---|
| 1 | 28.22 | 28.32 | 27.98 | 28.17 | 65.9 | 80.3 | 92 | 93 | 104 | 96.22 | 21.4 |
| 2 | 21.25 | 21.09 | 21.00 | 21.11 | 74.5 | 83.9 | 91 | 91 | 104 | 95.23 | 33.4 |
| 3 | 16.88 | 16.59 | 16.50 | 16.66 | 76.9 | 85.1 | 87 | 93 | 105 | 94.89 | 50.6 |
| 4 | 14.01 | 13.82 | 14.02 | 13.95 | 78.1 | 84.8 | 86 | 89 | 105 | 93.22 | 65.8 |
| 5 | 11.30 | 10.98 | 11.22 | 11.17 | 79.0 | 84.5 | 86 | 88 | 102 | 91.89 | 97.5 |
| 6 | 9.95 | 9.89 | 10.02 | 9.95 | 78.4 | 83.3 | 82 | 85 | 103 | 89.89 | 114.0 |
| 7 | 8.64 | 8.76 | 8.49 | 8.63 | 77.7 | 82.0 | 81 | 83 | 101 | 88.22 | 144.5 |
| 8 | 7.51 | 7.96 | 7.80 | 7.76 | 78.5 | 82.2 | 80 | 84 | 99 | 87.56 | 166.9 |

### For 0.14% of $Al_2O_3$ Nanofluid

| Sr. No. | $t_1$ (Sec) | $t_2$ (Sec) | $t_3$ (Sec) | $t_{avg}$ (Sec) | $T_{in}$ (°C) | $T_{out}$ (°C) | $T_1$ (°C) | $T_2$ (°C) | $T_3$ (°C) | $T_s$ (°C) | $\Delta P$ (Pa) |
|---|---|---|---|---|---|---|---|---|---|---|---|
| 1 | 30.20 | 30.02 | 30.02 | 30.08 | 63.9 | 82.8 | 90 | 91 | 102 | 94.19 | 22.7 |
| 2 | 22.00 | 21.06 | 22.05 | 21.70 | 69.9 | 83.0 | 89 | 89 | 102 | 93.20 | 38.4 |
| 3 | 18.00 | 18.03 | 17.85 | 17.96 | 73.7 | 84.8 | 85 | 91 | 103 | 92.86 | 52.9 |
| 4 | 14.03 | 14.52 | 13.66 | 14.07 | 74.7 | 83.4 | 84 | 87 | 103 | 91.19 | 78.8 |
| 5 | 11.22 | 10.88 | 11.02 | 11.04 | 75.5 | 82.8 | 84 | 86 | 100 | 89.85 | 121.6 |
| 6 | 10.05 | 10.03 | 9.86 | 9.98 | 75.3 | 81.6 | 80 | 83 | 101 | 87.86 | 138.2 |
| 7 | 9.03 | 9.06 | 8.96 | 9.02 | 74.6 | 80.2 | 79 | 81 | 99 | 86.19 | 161.6 |
| 8 | 8.00 | 7.63 | 8.03 | 7.89 | 74.6 | 79.8 | 78 | 82 | 97 | 85.52 | 197.1 |

### For 0.21% of $Al_2O_3$ Nanofluid

| Sr. No. | $t_1$ (Sec) | $t_2$ (Sec) | $t_3$ (Sec) | $t_{avg}$ (Sec) | $T_{in}$ (°C) | $T_{out}$ (°C) | $T_1$ (°C) | $T_2$ (°C) | $T_3$ (°C) | $T_s$ (°C) | $\Delta P$ (Pa) |
|---|---|---|---|---|---|---|---|---|---|---|---|
| 1 | 31.20 | 31.22 | 30.99 | 31.14 | 70.8 | 89.7 | 90 | 92 | 104 | 95.20 | 25.1 |
| 2 | 22.03 | 22.21 | 22.00 | 22.08 | 71.8 | 87.6 | 89 | 89 | 103 | 93.51 | 45.1 |
| 3 | 17.89 | 18.02 | 18.05 | 17.99 | 78.3 | 89.1 | 88 | 90 | 104 | 93.87 | 61.1 |
| 4 | 14.00 | 14.09 | 13.99 | 14.03 | 77.5 | 86.9 | 85 | 87 | 104 | 91.85 | 94.1 |
| 5 | 12.03 | 12.03 | 11.88 | 11.98 | 80.1 | 87.2 | 85 | 87 | 102 | 91.20 | 120.9 |
| 6 | 10.50 | 10.54 | 10.33 | 10.46 | 78.3 | 84.8 | 83 | 82 | 102 | 88.86 | 149.9 |
| 7 | 9.00 | 9.22 | 9.18 | 9.13 | 77.3 | 83.1 | 82 | 81 | 99 | 87.19 | 190.0 |
| 8 | 8.23 | 7.89 | 8.67 | 8.26 | 77.0 | 82.3 | 79 | 82 | 98 | 86.19 | 214.3 |

## Measured Parameter of Single Twisted Tape

### For Water

| Sr. No. | $t_1$ (Sec) | $t_2$ (Sec) | $t_3$ (Sec) | $t_{avg}$ (Sec) | $T_{in}$ ($^0$C) | $T_{out}$ ($^0$C) | $T_1$ ($^0$C) | $T_2$ ($^0$C) | $T_3$ ($^0$C) | $T_s$ ($^0$C) | $\Delta P$ (Pa) |
|---|---|---|---|---|---|---|---|---|---|---|---|
| 1 | 25.22 | 25.32 | 25.01 | 25.18 | 63.0 | 76.6 | 92 | 94 | 105 | 96.88 | 39.7 |
| 2 | 18.69 | 18.88 | 18.70 | 18.76 | 69.6 | 80.4 | 91 | 93 | 105 | 96.20 | 62.7 |
| 3 | 15.60 | 15.70 | 15.89 | 15.73 | 73.3 | 82.3 | 89 | 93 | 106 | 95.87 | 85.3 |
| 4 | 12.80 | 12.34 | 12.55 | 12.56 | 74.1 | 81.6 | 87 | 90 | 105 | 93.86 | 127.8 |
| 5 | 10.35 | 10.37 | 10.22 | 10.31 | 76.2 | 82.4 | 86 | 89 | 104 | 92.86 | 181.5 |
| 6 | 9.20 | 9.20 | 9.09 | 9.16 | 76.3 | 81.7 | 85 | 86 | 103 | 91.20 | 224.7 |
| 7 | 8.00 | 8.34 | 7.98 | 8.11 | 74.9 | 80.0 | 83 | 84 | 101 | 89.19 | 280.5 |
| 8 | 7.40 | 7.20 | 7.22 | 7.27 | 75.1 | 79.8 | 82 | 85 | 99 | 88.52 | 343.1 |

### For 0.07% of $Al_2O_3$ Nanofluid

| Sr. No. | $t_1$ (Sec) | $t_2$ (Sec) | $t_3$ (Sec) | $t_{avg}$ (Sec) | $T_{in}$ ($^0$C) | $T_{out}$ ($^0$C) | $T_1$ ($^0$C) | $T_2$ ($^0$C) | $T_3$ ($^0$C) | $T_s$ ($^0$C) | $\Delta P$ (Pa) |
|---|---|---|---|---|---|---|---|---|---|---|---|
| 1 | 28.36 | 28.03 | 27.96 | 28.12 | 64.8 | 81.4 | 94 | 97 | 108 | 99.67 | 39.8 |
| 2 | 21.22 | 21.52 | 21.00 | 21.25 | 71.8 | 84.4 | 94 | 98 | 104 | 98.67 | 61.0 |
| 3 | 16.95 | 16.66 | 16.99 | 16.87 | 75.2 | 85.3 | 92 | 98 | 109 | 99.67 | 92.7 |
| 4 | 13.89 | 13.85 | 14.02 | 13.92 | 76.7 | 85.0 | 88 | 89 | 107 | 94.67 | 130.3 |
| 5 | 11.30 | 10.96 | 11.30 | 11.19 | 78.3 | 84.8 | 90 | 91 | 105 | 95.33 | 193.4 |
| 6 | 9.96 | 9.97 | 10.05 | 9.99 | 77.2 | 83.4 | 87 | 90 | 105 | 94.00 | 237.2 |
| 7 | 9.00 | 9.02 | 7.85 | 8.62 | 77.2 | 82.2 | 85 | 85 | 103 | 91.00 | 311.6 |
| 8 | 7.50 | 7.96 | 7.80 | 7.75 | 77.5 | 82.1 | 84 | 86 | 100 | 90.00 | 379.7 |

### For 0.14% of $Al_2O_3$ Nanofluid

| Sr. No. | $t_1$ (Sec) | $t_2$ (Sec) | $t_3$ (Sec) | $t_{avg}$ (Sec) | $T_{in}$ ($^0$C) | $T_{out}$ ($^0$C) | $T_1$ ($^0$C) | $T_2$ ($^0$C) | $T_3$ ($^0$C) | $T_s$ ($^0$C) | $\Delta P$ (Pa) |
|---|---|---|---|---|---|---|---|---|---|---|---|
| 1 | 30.20 | 30.02 | 30.02 | 30.08 | 64.0 | 81.9 | 95 | 98 | 108 | 100.33 | 36.0 |
| 2 | 22.00 | 21.06 | 22.05 | 21.70 | 70.8 | 84.0 | 94 | 97 | 105 | 98.67 | 60.7 |
| 3 | 18.00 | 18.03 | 17.85 | 17.96 | 73.5 | 84.7 | 93 | 99 | 108 | 100.00 | 85.0 |
| 4 | 14.03 | 14.52 | 13.66 | 14.07 | 72.0 | 84.1 | 88 | 91 | 108 | 95.67 | 156.0 |
| 5 | 11.22 | 10.88 | 11.02 | 11.04 | 74.0 | 83.8 | 91 | 91 | 106 | 96.00 | 243.1 |
| 6 | 10.05 | 10.03 | 9.86 | 9.98 | 76.0 | 82.4 | 87 | 91 | 107 | 95.00 | 247.7 |
| 7 | 9.03 | 9.06 | 8.96 | 9.02 | 75.5 | 81.4 | 84 | 86 | 105 | 91.67 | 297.1 |
| 8 | 8.00 | 7.63 | 8.03 | 7.89 | 75.9 | 81.0 | 85 | 86 | 101 | 90.67 | 382.8 |

### For 0.21% of $Al_2O_3$ Nanofluid

| Sr. No. | $t_1$ (Sec) | $t_2$ (Sec) | $t_3$ (Sec) | $t_{avg}$ (Sec) | $T_{in}$ ($^0$C) | $T_{out}$ ($^0$C) | $T_1$ ($^0$C) | $T_2$ ($^0$C) | $T_3$ ($^0$C) | $T_s$ ($^0$C) | $\Delta P$ (Pa) |
|---|---|---|---|---|---|---|---|---|---|---|---|
| 1 | 31.20 | 31.22 | 30.99 | 31.14 | 67.0 | 90.4 | 95 | 99 | 108 | 100.67 | 47.2 |
| 2 | 22.03 | 22.21 | 22.00 | 22.08 | 73.0 | 90.0 | 95 | 97 | 105 | 99.00 | 82.5 |
| 3 | 17.89 | 18.02 | 18.05 | 17.99 | 76.0 | 90.7 | 92 | 100 | 110 | 100.67 | 119.3 |
| 4 | 14.00 | 14.09 | 13.99 | 14.03 | 80.0 | 89.1 | 89 | 91 | 110 | 96.67 | 188.4 |
| 5 | 12.03 | 12.03 | 11.88 | 11.98 | 81.5 | 89.3 | 92 | 93 | 106 | 97.00 | 247.9 |
| 6 | 10.50 | 10.54 | 10.33 | 10.46 | 80.2 | 87.0 | 89 | 90 | 109 | 96.00 | 318.9 |
| 7 | 9.00 | 9.22 | 9.18 | 9.13 | 79.7 | 85.6 | 86 | 87 | 108 | 93.67 | 409.5 |
| 8 | 8.23 | 7.89 | 8.67 | 8.26 | 79.5 | 84.9 | 87 | 86 | 103 | 92.00 | 493.7 |

## Measured Parameter of Dual Co-Tapes

### For Water

| Sr. No. | $t_1$ (Sec) | $t_2$ (Sec) | $t_3$ (Sec) | $t_{avg}$ (Sec) | $T_{in}$ ($^0$C) | $T_{out}$ ($^0$C) | $T_1$ ($^0$C) | $T_2$ ($^0$C) | $T_3$ ($^0$C) | $T_s$ ($^0$C) | $\Delta P$ (Pa) |
|---|---|---|---|---|---|---|---|---|---|---|---|
| 1 | 28.64 | 27.75 | 27.40 | 27.93 | 60.0 | 75.4 | 91 | 95 | 104 | 96.54 | 0.2 |
| 2 | 21.22 | 21.74 | 21.21 | 21.39 | 69.0 | 80.6 | 90 | 94 | 105 | 96.21 | 0.2 |
| 3 | 17.29 | 16.16 | 17.50 | 16.98 | 73.6 | 83.2 | 88 | 92 | 110 | 96.54 | 0.2 |
| 4 | 14.03 | 13.71 | 13.74 | 13.83 | 73.8 | 81.7 | 86 | 89 | 106 | 93.54 | 0.2 |
| 5 | 11.53 | 10.63 | 11.64 | 11.27 | 75.6 | 82.0 | 85 | 88 | 103 | 91.87 | 0.2 |
| 6 | 10.06 | 9.87 | 9.85 | 9.93 | 76.8 | 82.5 | 85 | 88 | 102 | 91.54 | 0.2 |
| 7 | 9.18 | 8.75 | 8.09 | 8.67 | 76.3 | 81.4 | 82 | 86 | 102 | 89.87 | 0.2 |
| 8 | 7.58 | 7.88 | 7.64 | 7.70 | 75.3 | 79.9 | 81 | 83 | 100 | 87.86 | 0.2 |

### For 0.07% of $Al_2O_3$ Nanofluid

| Sr. No. | $t_1$ (Sec) | $t_2$ (Sec) | $t_3$ (Sec) | $t_{avg}$ (Sec) | $T_{in}$ ($^0$C) | $T_{out}$ ($^0$C) | $T_1$ ($^0$C) | $T_2$ ($^0$C) | $T_3$ ($^0$C) | $T_s$ ($^0$C) | $\Delta P$ (Pa) |
|---|---|---|---|---|---|---|---|---|---|---|---|
| 1 | 30.50 | 29.72 | 29.42 | 29.88 | 69.0 | 84.0 | 93 | 98 | 107 | 99.33 | 0.2 |
| 2 | 22.00 | 21.27 | 22.27 | 21.85 | 74.6 | 86.4 | 93 | 99 | 104 | 98.67 | 0.2 |
| 3 | 18.36 | 17.49 | 18.39 | 18.08 | 77.2 | 86.8 | 91 | 97 | 113 | 100.33 | 0.2 |
| 4 | 14.17 | 14.37 | 13.39 | 13.98 | 78.2 | 86.2 | 87 | 88 | 108 | 94.33 | 0.2 |
| 5 | 11.44 | 10.55 | 11.35 | 11.12 | 79.3 | 85.6 | 89 | 90 | 104 | 94.33 | 0.2 |
| 6 | 10.15 | 9.93 | 9.66 | 9.91 | 78.5 | 84.1 | 87 | 92 | 104 | 94.33 | 0.2 |
| 7 | 9.21 | 8.79 | 9.23 | 9.08 | 78.0 | 82.8 | 84 | 87 | 104 | 91.67 | 0.2 |
| 8 | 8.08 | 7.55 | 7.87 | 7.83 | 77.9 | 82.4 | 83 | 84 | 101 | 89.33 | 0.2 |

### For 0.14% of $Al_2O_3$ Nanofluid

| Sr. No. | $t_1$ (Sec) | $t_2$ (Sec) | $t_3$ (Sec) | $t_{avg}$ (Sec) | $T_{in}$ ($^0$C) | $T_{out}$ ($^0$C) | $T_1$ ($^0$C) | $T_2$ ($^0$C) | $T_3$ ($^0$C) | $T_s$ ($^0$C) | $\Delta P$ (Pa) |
|---|---|---|---|---|---|---|---|---|---|---|---|
| 1 | 30.20 | 30.02 | 30.02 | 30.08 | 69.0 | 88.0 | 94 | 99 | 107 | 100.00 | 0.2 |
| 2 | 22.00 | 21.06 | 22.05 | 21.70 | 74.9 | 88.2 | 93 | 98 | 105 | 98.67 | 0.2 |
| 3 | 18.00 | 18.03 | 17.85 | 17.96 | 77.0 | 88.3 | 92 | 98 | 112 | 100.67 | 0.2 |
| 4 | 14.03 | 14.52 | 13.66 | 14.07 | 78.0 | 87.0 | 87 | 90 | 109 | 95.33 | 0.2 |
| 5 | 11.22 | 10.88 | 11.02 | 11.04 | 78.7 | 86.0 | 90 | 90 | 105 | 95.00 | 0.2 |
| 6 | 10.05 | 10.03 | 9.86 | 9.98 | 77.1 | 84.3 | 87 | 93 | 106 | 95.33 | 0.2 |
| 7 | 9.03 | 9.06 | 8.96 | 9.02 | 76.8 | 82.7 | 83 | 88 | 106 | 92.33 | 0.2 |
| 8 | 8.00 | 7.63 | 8.03 | 7.89 | 77.0 | 82.3 | 84 | 84 | 102 | 90.00 | 0.2 |

### For 0.21% of $Al_2O_3$ Nanofluid

| Sr. No. | $t_1$ (Sec) | $t_2$ (Sec) | $t_3$ (Sec) | $t_{avg}$ (Sec) | $T_{in}$ ($^0$C) | $T_{out}$ ($^0$C) | $T_1$ ($^0$C) | $T_2$ ($^0$C) | $T_3$ ($^0$C) | $T_s$ ($^0$C) | $\Delta P$ (Pa) |
|---|---|---|---|---|---|---|---|---|---|---|---|
| 1 | 31.51 | 30.91 | 30.37 | 30.93 | 73.4 | 93.6 | 94 | 100 | 107 | 100.33 | 0.2 |
| 2 | 22.03 | 22.43 | 22.22 | 22.23 | 77.4 | 91.4 | 94 | 98 | 105 | 99.00 | 0.2 |
| 3 | 18.25 | 17.48 | 18.59 | 18.11 | 79.5 | 91.9 | 91 | 99 | 114 | 101.33 | 0.2 |
| 4 | 14.14 | 13.95 | 13.71 | 13.93 | 80.8 | 89.9 | 88 | 90 | 111 | 96.33 | 0.2 |
| 5 | 12.27 | 11.67 | 12.24 | 12.06 | 81.9 | 89.6 | 91 | 92 | 105 | 96.00 | 0.2 |
| 6 | 10.61 | 10.43 | 10.12 | 10.39 | 80.7 | 87.4 | 89 | 92 | 108 | 96.33 | 0.2 |
| 7 | 9.18 | 8.94 | 9.46 | 9.19 | 79.5 | 85.4 | 85 | 89 | 109 | 94.33 | 0.2 |
| 8 | 8.31 | 7.81 | 8.50 | 8.21 | 79.3 | 84.7 | 86 | 84 | 104 | 91.33 | 0.2 |

## Measured Parameter of Dual Counter Tapes

### For Water

| Sr. No. | $t_1$ (Sec) | $t_2$ (Sec) | $t_3$ (Sec) | $t_{avg}$ (Sec) | $T_{in}$ (°C) | $T_{out}$ (°C) | $T_1$ (°C) | $T_2$ (°C) | $T_3$ (°C) | $T_s$ (°C) | $\Delta P$ (Pa) |
|---|---|---|---|---|---|---|---|---|---|---|---|
| 1 | 25.68 | 25.88 | 25.77 | 25.78 | 63.2 | 78.1 | 90 | 94 | 105 | 96.20 | 81.5 |
| 2 | 18.77 | 18.66 | 18.56 | 18.66 | 70.9 | 82.0 | 91 | 93 | 104 | 95.86 | 143.3 |
| 3 | 15.01 | 15.09 | 14.89 | 15.00 | 75.1 | 84.1 | 87 | 90 | 112 | 96.20 | 212.2 |
| 4 | 12.89 | 12.99 | 12.70 | 12.86 | 75.0 | 82.9 | 85 | 90 | 106 | 93.53 | 276.0 |
| 5 | 10.50 | 10.67 | 10.23 | 10.47 | 75.4 | 81.8 | 84 | 87 | 102 | 90.86 | 398.8 |
| 6 | 9.44 | 9.76 | 9.02 | 9.41 | 77.3 | 83.0 | 85 | 88 | 101 | 91.19 | 482.3 |
| 7 | 8.04 | 8.66 | 8.01 | 8.24 | 76.4 | 81.7 | 83 | 85 | 101 | 89.52 | 614.7 |
| 8 | 7.44 | 7.59 | 7.08 | 7.37 | 75.6 | 80.4 | 82 | 82 | 100 | 87.85 | 755.9 |

### For 0.07% of $Al_2O_3$ Nanofluid

| Sr. No. | $t_1$ (Sec) | $t_2$ (Sec) | $t_3$ (Sec) | $t_{avg}$ (Sec) | $T_{in}$ (°C) | $T_{out}$ (°C) | $T_1$ (°C) | $T_2$ (°C) | $T_3$ (°C) | $T_s$ (°C) | $\Delta P$ (Pa) |
|---|---|---|---|---|---|---|---|---|---|---|---|
| 1 | 28.93 | 27.47 | 26.85 | 27.75 | 68.0 | 85.4 | 92 | 97 | 108 | 99.00 | 86.9 |
| 2 | 21.22 | 21.95 | 21.42 | 21.53 | 74.0 | 86.5 | 94 | 98 | 103 | 98.33 | 133.2 |
| 3 | 17.63 | 15.68 | 18.02 | 17.11 | 77.0 | 87.2 | 90 | 95 | 115 | 100.00 | 201.9 |
| 4 | 14.17 | 13.57 | 13.46 | 13.74 | 78.2 | 86.5 | 86 | 89 | 108 | 94.33 | 300.0 |
| 5 | 11.76 | 10.31 | 11.99 | 11.35 | 79.0 | 86.0 | 88 | 89 | 103 | 93.33 | 420.9 |
| 6 | 10.16 | 9.77 | 9.65 | 9.86 | 78.5 | 84.4 | 87 | 92 | 103 | 94.00 | 546.0 |
| 7 | 9.36 | 8.49 | 8.33 | 8.73 | 77.8 | 83.2 | 85 | 86 | 103 | 91.33 | 682.0 |
| 8 | 7.65 | 7.80 | 7.49 | 7.65 | 78.2 | 82.8 | 84 | 83 | 101 | 89.33 | 874.6 |

### For 0.14% of $Al_2O_3$ Nanofluid

| Sr. No. | $t_1$ (Sec) | $t_2$ (Sec) | $t_3$ (Sec) | $t_{avg}$ (Sec) | $T_{in}$ (°C) | $T_{out}$ (°C) | $T_1$ (°C) | $T_2$ (°C) | $T_3$ (°C) | $T_s$ (°C) | $\Delta P$ (Pa) |
|---|---|---|---|---|---|---|---|---|---|---|---|
| 1 | 30.81 | 29.42 | 28.83 | 29.69 | 71.8 | 89.9 | 93 | 98 | 108 | 99.67 | 90.9 |
| 2 | 22.00 | 21.48 | 22.49 | 21.99 | 75.5 | 88.9 | 94 | 97 | 104 | 98.33 | 153.3 |
| 3 | 18.73 | 16.96 | 18.94 | 18.21 | 77.9 | 89.3 | 91 | 96 | 114 | 100.33 | 214.5 |
| 4 | 14.31 | 14.23 | 13.12 | 13.89 | 78.6 | 87.6 | 86 | 91 | 109 | 95.33 | 353.3 |
| 5 | 11.67 | 10.24 | 11.69 | 11.20 | 79.6 | 86.8 | 89 | 89 | 104 | 94.00 | 521.1 |
| 6 | 10.25 | 9.83 | 9.47 | 9.85 | 77.2 | 84.8 | 87 | 93 | 105 | 95.00 | 660.1 |
| 7 | 9.39 | 8.52 | 9.51 | 9.14 | 78.0 | 83.8 | 84 | 87 | 105 | 92.00 | 750.3 |
| 8 | 8.16 | 7.48 | 7.71 | 7.78 | 77.0 | 82.9 | 85 | 83 | 102 | 90.00 | 1020.4 |

### For 0.21% of $Al_2O_3$ Nanofluid

| Sr. No. | $t_1$ (Sec) | $t_2$ (Sec) | $t_3$ (Sec) | $t_{avg}$ (Sec) | $T_{in}$ (°C) | $T_{out}$ (°C) | $T_1$ (°C) | $T_2$ (°C) | $T_3$ (°C) | $T_s$ (°C) | $\Delta P$ (Pa) |
|---|---|---|---|---|---|---|---|---|---|---|---|
| 1 | 31.83 | 30.60 | 29.76 | 30.73 | 74.4 | 94.0 | 93 | 99 | 108 | 100.00 | 99.4 |
| 2 | 22.03 | 22.66 | 22.44 | 22.38 | 77.7 | 92.3 | 95 | 97 | 104 | 98.67 | 173.6 |
| 3 | 18.61 | 16.96 | 19.15 | 18.24 | 80.9 | 92.5 | 90 | 97 | 116 | 101.00 | 250.9 |
| 4 | 14.28 | 13.81 | 13.44 | 13.84 | 81.1 | 90.5 | 87 | 91 | 111 | 96.33 | 417.8 |
| 5 | 12.52 | 11.32 | 12.60 | 12.15 | 82.2 | 90.0 | 90 | 91 | 104 | 95.00 | 521.1 |
| 6 | 10.71 | 10.33 | 9.92 | 10.32 | 81.5 | 88.0 | 89 | 92 | 107 | 96.00 | 707.9 |
| 7 | 9.36 | 8.68 | 9.74 | 9.26 | 79.0 | 86.0 | 86 | 88 | 108 | 94.00 | 861.4 |
| 8 | 8.40 | 7.73 | 8.33 | 8.15 | 79.9 | 85.2 | 87 | 83 | 104 | 91.33 | 1097.7 |

## Measured Parameter of Triple Co-Tapes
### For Water

| Sr. No. | $t_1$ (Sec) | $t_2$ (Sec) | $t_3$ (Sec) | $t_{avg}$ (Sec) | $T_{in}$ (°C) | $T_{out}$ (°C) | $T_1$ (°C) | $T_2$ (°C) | $T_3$ (°C) | $T_s$ (°C) | $\Delta P$ (Pa) |
|---|---|---|---|---|---|---|---|---|---|---|---|
| 1 | 26.00 | 26.03 | 25.88 | 25.97 | 66.0 | 81.4 | 91 | 93 | 106 | 96.53 | 111.9 |
| 2 | 18.99 | 19.05 | 19.01 | 19.02 | 72.2 | 83.8 | 90 | 92 | 105 | 95.53 | 192.4 |
| 3 | 15.00 | 15.08 | 14.99 | 15.02 | 76.2 | 85.3 | 88 | 91 | 108 | 95.53 | 294.8 |
| 4 | 13.02 | 12.99 | 13.44 | 13.15 | 76.7 | 85.0 | 85 | 93 | 104 | 93.86 | 367.7 |
| 5 | 10.66 | 10.88 | 10.89 | 10.81 | 76.9 | 83.6 | 85 | 88 | 101 | 91.19 | 521.4 |
| 6 | 9.43 | 9.43 | 9.02 | 9.29 | 79.4 | 85.2 | 86 | 89 | 102 | 92.19 | 688.4 |
| 7 | 8.06 | 8.56 | 8.99 | 8.54 | 77.0 | 82.6 | 83 | 84 | 101 | 89.18 | 797.9 |
| 8 | 7.43 | 7.44 | 7.60 | 7.49 | 77.2 | 82.2 | 83 | 84 | 99 | 88.51 | 1019.9 |

### For 0.07% of $Al_2O_3$ Nanofluid

| Sr. No. | $t_1$ (Sec) | $t_2$ (Sec) | $t_3$ (Sec) | $t_{avg}$ (Sec) | $T_{in}$ (°C) | $T_{out}$ (°C) | $T_1$ (°C) | $T_2$ (°C) | $T_3$ (°C) | $T_s$ (°C) | $\Delta P$ (Pa) |
|---|---|---|---|---|---|---|---|---|---|---|---|
| 1 | 29.22 | 27.75 | 27.12 | 28.03 | 67.0 | 86.3 | 93 | 96 | 109 | 99.33 | 118.7 |
| 2 | 21.43 | 22.17 | 21.64 | 21.75 | 74.8 | 87.7 | 93 | 97 | 104 | 98.00 | 182.0 |
| 3 | 17.81 | 15.83 | 18.20 | 17.28 | 77.2 | 88.0 | 91 | 96 | 111 | 99.33 | 276.0 |
| 4 | 14.31 | 13.71 | 13.60 | 13.87 | 78.2 | 87.3 | 86 | 92 | 106 | 94.67 | 410.0 |
| 5 | 11.87 | 10.42 | 12.11 | 11.47 | 79.3 | 86.6 | 89 | 90 | 102 | 93.67 | 575.3 |
| 6 | 10.26 | 9.87 | 9.75 | 9.96 | 78.3 | 85.1 | 88 | 93 | 104 | 95.00 | 746.4 |
| 7 | 9.46 | 8.57 | 8.41 | 8.81 | 78.2 | 83.8 | 85 | 85 | 103 | 91.00 | 932.3 |
| 8 | 7.73 | 7.88 | 7.57 | 7.72 | 78.2 | 83.3 | 85 | 85 | 100 | 90.00 | 1195.5 |

### For 0.14% of $Al_2O_3$ Nanofluid

| Sr. No. | $t_1$ (Sec) | $t_2$ (Sec) | $t_3$ (Sec) | $t_{avg}$ (Sec) | $T_{in}$ (°C) | $T_{out}$ (°C) | $T_1$ (°C) | $T_2$ (°C) | $T_3$ (°C) | $T_s$ (°C) | $\Delta P$ (Pa) |
|---|---|---|---|---|---|---|---|---|---|---|---|
| 1 | 31.12 | 29.72 | 29.12 | 29.98 | 71.6 | 89.7 | 94 | 97 | 109 | 100.00 | 125.8 |
| 2 | 22.22 | 21.70 | 22.72 | 22.21 | 73.0 | 88.6 | 93 | 96 | 105 | 98.00 | 212.1 |
| 3 | 18.91 | 17.13 | 19.13 | 18.39 | 78.0 | 88.9 | 92 | 97 | 110 | 99.67 | 296.6 |
| 4 | 14.46 | 14.37 | 13.25 | 14.03 | 78.0 | 87.4 | 86 | 94 | 107 | 95.67 | 488.7 |
| 5 | 11.79 | 10.34 | 11.81 | 11.31 | 78.5 | 86.2 | 90 | 90 | 103 | 94.33 | 720.8 |
| 6 | 10.35 | 9.93 | 9.56 | 9.95 | 78.3 | 84.7 | 88 | 94 | 106 | 96.00 | 912.6 |
| 7 | 9.49 | 8.61 | 9.60 | 9.23 | 77.2 | 83.1 | 84 | 86 | 105 | 91.67 | 1037.1 |
| 8 | 8.24 | 7.55 | 7.79 | 7.86 | 77.3 | 82.6 | 86 | 85 | 101 | 90.67 | 1410.5 |

### For 0.21% of $Al_2O_3$ Nanofluid

| Sr. No. | $t_1$ (Sec) | $t_2$ (Sec) | $t_3$ (Sec) | $t_{avg}$ (Sec) | $T_{in}$ (°C) | $T_{out}$ (°C) | $T_1$ (°C) | $T_2$ (°C) | $T_3$ (°C) | $T_s$ (°C) | $\Delta P$ (Pa) |
|---|---|---|---|---|---|---|---|---|---|---|---|
| 1 | 32.15 | 30.90 | 30.06 | 31.04 | 74.3 | 94.4 | 94 | 98 | 109 | 100.33 | 138.8 |
| 2 | 22.25 | 22.88 | 22.67 | 22.60 | 78.1 | 92.3 | 94 | 96 | 105 | 98.33 | 242.6 |
| 3 | 18.80 | 17.12 | 19.34 | 18.42 | 80.5 | 92.1 | 91 | 98 | 112 | 100.33 | 350.3 |
| 4 | 14.42 | 13.95 | 13.57 | 13.98 | 81.1 | 90.1 | 87 | 94 | 109 | 96.67 | 583.6 |
| 5 | 12.64 | 11.43 | 12.73 | 12.27 | 82.0 | 89.7 | 91 | 92 | 103 | 95.33 | 727.5 |
| 6 | 10.82 | 10.43 | 10.02 | 10.42 | 80.5 | 87.3 | 90 | 93 | 108 | 97.00 | 988.6 |
| 7 | 9.46 | 8.76 | 9.84 | 9.35 | 79.5 | 85.4 | 86 | 87 | 108 | 93.67 | 1203.9 |
| 8 | 8.48 | 7.81 | 8.41 | 8.23 | 79.2 | 84.5 | 88 | 85 | 103 | 92.00 | 1532.9 |

## Measured Parameter of Triple Counter Tapes
### For Water

| Sr. No. | $t_1$ (Sec) | $t_2$ (Sec) | $t_3$ (Sec) | $t_{avg}$ (Sec) | $T_{in}$ (°C) | $T_{out}$ (°C) | $T_1$ (°C) | $T_2$ (°C) | $T_3$ (°C) | $T_s$ (°C) | $\Delta P$ (Pa) |
|---|---|---|---|---|---|---|---|---|---|---|---|
| 1 | 26.43 | 26.45 | 26.33 | 26.40 | 66.3 | 81.9 | 92 | 91 | 105 | 95.86 | 113.7 |
| 2 | 19.45 | 19.43 | 19.09 | 19.32 | 73.0 | 85.0 | 89 | 93 | 106 | 95.86 | 195.5 |
| 3 | 15.23 | 14.98 | 15.08 | 15.10 | 77.1 | 86.4 | 89 | 92 | 107 | 95.86 | 306.3 |
| 4 | 13.40 | 13.40 | 13.11 | 13.30 | 76.6 | 85.2 | 84 | 92 | 105 | 93.52 | 377.3 |
| 5 | 10.55 | 10.67 | 10.45 | 10.56 | 77.2 | 84.0 | 84 | 87 | 103 | 91.19 | 573.6 |
| 6 | 9.44 | 9.67 | 9.33 | 9.48 | 78.4 | 84.6 | 85 | 88 | 101 | 91.18 | 694.6 |
| 7 | 8.07 | 8.22 | 8.67 | 8.32 | 76.8 | 82.5 | 82 | 83 | 102 | 88.84 | 882.0 |
| 8 | 7.67 | 7.32 | 7.49 | 7.49 | 77.2 | 82.3 | 81 | 84 | 100 | 88.18 | 1070.0 |

### For 0.07% of $Al_2O_3$ Nanofluid

| Sr. No. | $t_1$ (Sec) | $t_2$ (Sec) | $t_3$ (Sec) | $t_{avg}$ (Sec) | $T_{in}$ (°C) | $T_{out}$ (°C) | $T_1$ (°C) | $T_2$ (°C) | $T_3$ (°C) | $T_s$ (°C) | $\Delta P$ (Pa) |
|---|---|---|---|---|---|---|---|---|---|---|---|
| 1 | 28.93 | 28.03 | 27.67 | 28.21 | 71.2 | 88.0 | 94 | 94 | 108 | 98.67 | 123.3 |
| 2 | 21.43 | 21.95 | 21.42 | 21.60 | 76.0 | 89.1 | 92 | 98 | 105 | 98.33 | 194.0 |
| 3 | 17.46 | 16.32 | 17.67 | 17.15 | 79.4 | 89.3 | 92 | 97 | 110 | 99.67 | 294.6 |
| 4 | 14.17 | 13.85 | 13.88 | 13.96 | 80.5 | 88.4 | 85 | 91 | 107 | 94.33 | 425.2 |
| 5 | 11.64 | 10.74 | 11.76 | 11.38 | 81.0 | 87.6 | 88 | 89 | 104 | 93.67 | 613.8 |
| 6 | 10.16 | 9.97 | 9.95 | 10.03 | 80.0 | 86.0 | 87 | 92 | 103 | 94.00 | 774.0 |
| 7 | 9.27 | 8.84 | 8.17 | 8.76 | 79.2 | 84.5 | 84 | 84 | 104 | 90.67 | 991.9 |
| 8 | 7.65 | 7.96 | 7.72 | 7.78 | 79.4 | 84.1 | 83 | 85 | 101 | 89.67 | 1239.2 |

### For 0.14% of $Al_2O_3$ Nanofluid

| Sr. No. | $t_1$ (Sec) | $t_2$ (Sec) | $t_3$ (Sec) | $t_{avg}$ (Sec) | $T_{in}$ (°C) | $T_{out}$ (°C) | $T_1$ (°C) | $T_2$ (°C) | $T_3$ (°C) | $T_s$ (°C) | $\Delta P$ (Pa) |
|---|---|---|---|---|---|---|---|---|---|---|---|
| 1 | 30.81 | 30.02 | 29.71 | 30.18 | 70.0 | 90.8 | 95 | 95 | 108 | 99.33 | 130.1 |
| 2 | 22.22 | 21.48 | 22.49 | 22.07 | 78.0 | 90.5 | 92 | 97 | 106 | 98.33 | 225.1 |
| 3 | 18.54 | 17.66 | 18.57 | 18.26 | 78.7 | 90.1 | 93 | 98 | 109 | 100.00 | 315.3 |
| 4 | 14.31 | 14.52 | 13.52 | 14.12 | 79.0 | 88.1 | 85 | 93 | 108 | 95.33 | 505.6 |
| 5 | 11.56 | 10.66 | 11.46 | 11.23 | 79.5 | 87.1 | 89 | 89 | 105 | 94.33 | 766.9 |
| 6 | 10.25 | 10.03 | 9.76 | 10.01 | 78.2 | 85.3 | 87 | 93 | 105 | 95.00 | 944.7 |
| 7 | 9.30 | 8.88 | 9.32 | 9.17 | 78.0 | 83.9 | 83 | 85 | 106 | 91.33 | 1103.5 |
| 8 | 8.16 | 7.63 | 7.95 | 7.91 | 77.4 | 83.1 | 84 | 85 | 102 | 90.33 | 1460.8 |

### For 0.21% of $Al_2O_3$ Nanofluid

| Sr. No. | $t_1$ (Sec) | $t_2$ (Sec) | $t_3$ (Sec) | $t_{avg}$ (Sec) | $T_{in}$ (°C) | $T_{out}$ (°C) | $T_1$ (°C) | $T_2$ (°C) | $T_3$ (°C) | $T_s$ (°C) | $\Delta P$ (Pa) |
|---|---|---|---|---|---|---|---|---|---|---|---|
| 1 | 31.83 | 31.22 | 30.67 | 31.24 | 75.3 | 95.4 | 95 | 96 | 108 | 99.67 | 143.6 |
| 2 | 22.25 | 22.66 | 22.44 | 22.45 | 78.6 | 92.9 | 93 | 97 | 106 | 98.67 | 257.7 |
| 3 | 18.43 | 17.65 | 18.78 | 18.29 | 81.3 | 92.9 | 92 | 99 | 111 | 100.67 | 372.6 |
| 4 | 14.28 | 14.09 | 13.85 | 14.07 | 80.0 | 90.4 | 86 | 93 | 110 | 96.33 | 604.3 |
| 5 | 12.39 | 11.79 | 12.36 | 12.18 | 82.4 | 90.1 | 90 | 91 | 105 | 95.33 | 774.2 |
| 6 | 10.71 | 10.54 | 10.22 | 10.49 | 81.0 | 87.8 | 89 | 92 | 107 | 96.00 | 1023.9 |
| 7 | 9.27 | 9.03 | 9.55 | 9.28 | 79.6 | 86.0 | 85 | 86 | 109 | 93.33 | 1280.9 |
| 8 | 8.40 | 7.89 | 8.58 | 8.29 | 79.7 | 85.0 | 86 | 85 | 104 | 91.67 | 1586.8 |

## Measured Parameter of Quadruple Co-Tapes Swirl Flow Generators

### For Water

| Sr. No. | $t_1$ (Sec) | $t_2$ (Sec) | $t_3$ (Sec) | $t_{avg}$ (Sec) | $T_{in}$ (°C) | $T_{out}$ (°C) | $T_1$ (°C) | $T_2$ (°C) | $T_3$ (°C) | $T_s$ (°C) | $\Delta P$ (Pa) |
|---|---|---|---|---|---|---|---|---|---|---|---|
| 1 | 26.21 | 26.42 | 26.30 | 26.31 | 68.9 | 84.2 | 91 | 90 | 103 | 94.53 | 145.2 |
| 2 | 19.30 | 19.40 | 19.33 | 19.34 | 75.1 | 86.8 | 88 | 92 | 105 | 94.86 | 247.4 |
| 3 | 15.20 | 15.40 | 15.02 | 15.21 | 77.5 | 87.0 | 88 | 90 | 106 | 94.52 | 383.1 |
| 4 | 13.30 | 13.29 | 13.44 | 13.34 | 77.8 | 86.2 | 83 | 91 | 104 | 92.52 | 475.8 |
| 5 | 10.77 | 10.33 | 10.57 | 10.56 | 78.0 | 84.6 | 82 | 86 | 103 | 90.19 | 728.4 |
| 6 | 9.30 | 9.22 | 9.66 | 9.39 | 79.0 | 84.8 | 83 | 86 | 101 | 89.86 | 898.4 |
| 7 | 8.21 | 8.34 | 8.65 | 8.40 | 77.6 | 83.2 | 81 | 82 | 102 | 88.18 | 1097.9 |
| 8 | 7.55 | 7.65 | 7.45 | 7.55 | 77.9 | 82.8 | 81 | 83 | 99 | 87.52 | 1338.4 |

### For 0.07% of $Al_2O_3$ Nanofluid

| Sr. No. | $t_1$ (Sec) | $t_2$ (Sec) | $t_3$ (Sec) | $t_{avg}$ (Sec) | $T_{in}$ (°C) | $T_{out}$ (°C) | $T_1$ (°C) | $T_2$ (°C) | $T_3$ (°C) | $T_s$ (°C) | $\Delta P$ (Pa) |
|---|---|---|---|---|---|---|---|---|---|---|---|
| 1 | 29.51 | 28.59 | 28.23 | 28.77 | 66.0 | 88.6 | 93 | 93 | 106 | 97.33 | 150.1 |
| 2 | 21.86 | 22.39 | 21.85 | 22.03 | 74.6 | 89.5 | 91 | 97 | 104 | 97.33 | 236.3 |
| 3 | 17.81 | 16.65 | 18.03 | 17.50 | 76.0 | 89.6 | 91 | 95 | 109 | 98.33 | 359.1 |
| 4 | 14.45 | 14.13 | 14.15 | 14.24 | 78.2 | 88.5 | 84 | 90 | 106 | 93.33 | 518.6 |
| 5 | 11.87 | 10.95 | 11.99 | 11.61 | 79.3 | 87.7 | 86 | 88 | 104 | 92.67 | 748.3 |
| 6 | 10.36 | 10.17 | 10.15 | 10.23 | 77.9 | 86.0 | 85 | 90 | 103 | 92.67 | 944.1 |
| 7 | 9.46 | 9.01 | 8.33 | 8.93 | 78.0 | 84.6 | 83 | 83 | 104 | 90.00 | 1209.9 |
| 8 | 7.80 | 8.12 | 7.87 | 7.93 | 77.8 | 84.0 | 83 | 84 | 100 | 89.00 | 1511.5 |

### For 0.14% of $Al_2O_3$ Nanofluid

| Sr. No. | $t_1$ (Sec) | $t_2$ (Sec) | $t_3$ (Sec) | $t_{avg}$ (Sec) | $T_{in}$ (°C) | $T_{out}$ (°C) | $T_1$ (°C) | $T_2$ (°C) | $T_3$ (°C) | $T_s$ (°C) | $\Delta P$ (Pa) |
|---|---|---|---|---|---|---|---|---|---|---|---|
| 1 | 31.42 | 30.62 | 30.31 | 30.78 | 75.2 | 93.3 | 94 | 94 | 106 | 98.00 | 157.4 |
| 2 | 22.66 | 21.91 | 22.94 | 22.51 | 73.5 | 91.8 | 91 | 96 | 105 | 97.33 | 272.1 |
| 3 | 18.91 | 18.02 | 18.94 | 18.62 | 79.0 | 92.1 | 92 | 96 | 108 | 98.67 | 381.1 |
| 4 | 14.60 | 14.81 | 13.79 | 14.40 | 79.0 | 90.0 | 84 | 92 | 107 | 94.33 | 611.2 |
| 5 | 11.79 | 10.87 | 11.69 | 11.45 | 80.0 | 88.6 | 87 | 88 | 105 | 93.33 | 926.9 |
| 6 | 10.46 | 10.23 | 9.95 | 10.21 | 80.0 | 87.0 | 85 | 91 | 105 | 93.67 | 1141.8 |
| 7 | 9.49 | 9.05 | 9.51 | 9.35 | 80.0 | 85.3 | 82 | 84 | 106 | 90.67 | 1334.4 |
| 8 | 8.32 | 7.78 | 8.11 | 8.07 | 78.5 | 84.6 | 84 | 84 | 101 | 89.67 | 1765.6 |

### For 0.21% of $Al_2O_3$ Nanofluid

| Sr. No. | $t_1$ (Sec) | $t_2$ (Sec) | $t_3$ (Sec) | $t_{avg}$ (Sec) | $T_{in}$ (°C) | $T_{out}$ (°C) | $T_1$ (°C) | $T_2$ (°C) | $T_3$ (°C) | $T_s$ (°C) | $\Delta P$ (Pa) |
|---|---|---|---|---|---|---|---|---|---|---|---|
| 1 | 32.46 | 31.84 | 31.29 | 31.86 | 77.0 | 97.5 | 94 | 95 | 106 | 98.33 | 171.9 |
| 2 | 22.70 | 23.11 | 22.89 | 22.90 | 80.8 | 95.0 | 92 | 96 | 105 | 97.67 | 308.3 |
| 3 | 18.80 | 18.01 | 19.15 | 18.65 | 83.4 | 95.0 | 91 | 97 | 110 | 99.33 | 446.0 |
| 4 | 14.57 | 14.37 | 14.12 | 14.35 | 83.4 | 92.5 | 85 | 92 | 109 | 95.33 | 722.9 |
| 5 | 12.64 | 12.02 | 12.61 | 12.42 | 84.1 | 91.9 | 88 | 90 | 105 | 94.33 | 926.7 |
| 6 | 10.93 | 10.75 | 10.43 | 10.70 | 82.6 | 89.4 | 87 | 90 | 107 | 94.67 | 1225.5 |
| 7 | 9.46 | 9.21 | 9.74 | 9.47 | 80.0 | 87.2 | 84 | 85 | 109 | 92.67 | 1533.1 |
| 8 | 8.56 | 8.05 | 8.75 | 8.45 | 81.3 | 86.6 | 86 | 84 | 103 | 91.00 | 1899.3 |

## Measured Parameter of Quadruple Counter Tapes as Parallel Quadruple Swirl Flow Generators

### For Water

| Sr. No. | $t_1$ (Sec) | $t_2$ (Sec) | $t_3$ (Sec) | $t_{avg}$ (Sec) | $T_{in}$ ($^0$C) | $T_{out}$ ($^0$C) | $T_1$ ($^0$C) | $T_2$ ($^0$C) | $T_3$ ($^0$C) | $T_s$ ($^0$C) | $\Delta P$ (Pa) |
|---|---|---|---|---|---|---|---|---|---|---|---|
| 1 | 26.12 | 26.01 | 26.77 | 26.30 | 69.0 | 84.5 | 90 | 91 | 102 | 94.20 | 148.9 |
| 2 | 19.44 | 19.55 | 19.20 | 19.40 | 75.0 | 86.9 | 88 | 92 | 104 | 94.53 | 252.2 |
| 3 | 15.44 | 15.11 | 15.29 | 15.28 | 77.5 | 87.2 | 87 | 91 | 105 | 94.19 | 388.6 |
| 4 | 13.39 | 13.11 | 13.30 | 13.27 | 77.9 | 86.2 | 82 | 90 | 105 | 92.19 | 493.4 |
| 5 | 10.55 | 10.49 | 10.88 | 10.64 | 78.2 | 84.9 | 82 | 85 | 104 | 90.19 | 734.3 |
| 6 | 9.37 | 9.49 | 9.23 | 9.36 | 79.9 | 85.7 | 83 | 86 | 103 | 90.52 | 926.0 |
| 7 | 8.40 | 8.42 | 8.22 | 8.35 | 78.0 | 83.5 | 80 | 82 | 103 | 88.18 | 1140.7 |
| 8 | 7.81 | 7.66 | 7.45 | 7.64 | 78.7 | 83.7 | 81 | 83 | 101 | 88.18 | 1338.6 |

### For 0.07% of Al$_2$O$_3$ Nanofluid

| Sr. No. | $t_1$ (Sec) | $t_2$ (Sec) | $t_3$ (Sec) | $t_{avg}$ (Sec) | $T_{in}$ ($^0$C) | $T_{out}$ ($^0$C) | $T_1$ ($^0$C) | $T_2$ ($^0$C) | $T_3$ ($^0$C) | $T_s$ ($^0$C) | $\Delta P$ (Pa) |
|---|---|---|---|---|---|---|---|---|---|---|---|
| 1 | 29.80 | 28.30 | 27.66 | 28.59 | 70.1 | 91.0 | 92 | 94 | 105 | 97.00 | 155.5 |
| 2 | 21.86 | 22.62 | 22.07 | 22.18 | 78.6 | 91.3 | 91 | 97 | 103 | 97.00 | 238.6 |
| 3 | 18.17 | 16.15 | 18.57 | 17.63 | 80.0 | 91.2 | 90 | 96 | 108 | 98.00 | 361.4 |
| 4 | 14.60 | 13.98 | 13.87 | 14.15 | 81.0 | 89.8 | 83 | 89 | 107 | 93.00 | 537.4 |
| 5 | 12.11 | 10.62 | 12.35 | 11.70 | 81.3 | 88.8 | 86 | 87 | 105 | 92.67 | 754.1 |
| 6 | 10.47 | 10.07 | 9.94 | 10.16 | 80.5 | 87.0 | 85 | 90 | 105 | 93.33 | 980.5 |
| 7 | 9.65 | 8.74 | 8.58 | 8.99 | 80.3 | 85.4 | 82 | 83 | 105 | 90.00 | 1224.7 |
| 8 | 7.88 | 8.04 | 7.72 | 7.88 | 79.7 | 84.8 | 83 | 84 | 102 | 89.67 | 1571.4 |

### For 0.14% of Al$_2$O$_3$ Nanofluid

| Sr. No. | $t_1$ (Sec) | $t_2$ (Sec) | $t_3$ (Sec) | $t_{avg}$ (Sec) | $T_{in}$ ($^0$C) | $T_{out}$ ($^0$C) | $T_1$ ($^0$C) | $T_2$ ($^0$C) | $T_3$ ($^0$C) | $T_s$ ($^0$C) | $\Delta P$ (Pa) |
|---|---|---|---|---|---|---|---|---|---|---|---|
| 1 | 31.74 | 30.31 | 29.70 | 30.58 | 74.4 | 94.2 | 93 | 95 | 105 | 97.67 | 163.2 |
| 2 | 22.66 | 22.13 | 23.17 | 22.66 | 75.3 | 92.3 | 91 | 96 | 104 | 97.00 | 275.1 |
| 3 | 19.29 | 17.48 | 19.51 | 18.76 | 81.3 | 92.4 | 91 | 97 | 107 | 98.33 | 384.2 |
| 4 | 14.74 | 14.66 | 13.52 | 14.31 | 79.2 | 90.2 | 83 | 91 | 108 | 94.00 | 633.5 |
| 5 | 12.03 | 10.55 | 12.04 | 11.54 | 81.0 | 89.0 | 87 | 87 | 106 | 93.33 | 934.9 |
| 6 | 10.56 | 10.13 | 9.76 | 10.15 | 79.0 | 87.0 | 85 | 91 | 107 | 94.33 | 1184.9 |
| 7 | 9.68 | 8.78 | 9.79 | 9.42 | 79.0 | 85.3 | 81 | 84 | 107 | 90.67 | 1346.8 |
| 8 | 8.41 | 7.70 | 7.94 | 8.02 | 78.0 | 84.7 | 84 | 84 | 103 | 90.33 | 1832.5 |

### For 0.21% of Al$_2$O$_3$ Nanofluid

| Sr. No. | $t_1$ (Sec) | $t_2$ (Sec) | $t_3$ (Sec) | $t_{avg}$ (Sec) | $T_{in}$ ($^0$C) | $T_{out}$ ($^0$C) | $T_1$ ($^0$C) | $T_2$ ($^0$C) | $T_3$ ($^0$C) | $T_s$ ($^0$C) | $\Delta P$ (Pa) |
|---|---|---|---|---|---|---|---|---|---|---|---|
| 1 | 32.79 | 31.52 | 30.66 | 31.66 | 78.2 | 98.0 | 93 | 96 | 105 | 98.00 | 178.2 |
| 2 | 22.70 | 23.34 | 23.12 | 23.05 | 81.0 | 95.3 | 92 | 96 | 104 | 97.33 | 311.5 |
| 3 | 19.17 | 17.47 | 19.73 | 18.79 | 83.5 | 95.0 | 90 | 98 | 109 | 99.00 | 450.0 |
| 4 | 14.71 | 14.23 | 13.84 | 14.26 | 83.6 | 92.7 | 84 | 91 | 110 | 95.00 | 749.8 |
| 5 | 12.89 | 11.66 | 12.98 | 12.51 | 84.3 | 92.0 | 88 | 89 | 106 | 94.33 | 935.1 |
| 6 | 11.03 | 10.64 | 10.22 | 10.63 | 82.8 | 89.6 | 87 | 90 | 109 | 95.33 | 1272.5 |
| 7 | 9.65 | 8.94 | 10.03 | 9.54 | 81.6 | 87.6 | 83 | 85 | 110 | 92.67 | 1548.4 |
| 8 | 8.65 | 7.97 | 8.58 | 8.40 | 81.4 | 86.8 | 86 | 84 | 105 | 91.67 | 1971.6 |

## Measured Parameter of Quadruple Counter Tapes as Counter Quadruple Swirl Flow Generators

### For Water

| Sr. No. | $t_1$ (Sec) | $t_2$ (Sec) | $t_3$ (Sec) | $t_{avg}$ (Sec) | $T_{in}$ (°C) | $T_{out}$ (°C) | $T_1$ (°C) | $T_2$ (°C) | $T_3$ (°C) | $T_s$ (°C) | $\Delta P$ (Pa) |
|---|---|---|---|---|---|---|---|---|---|---|---|
| 1 | 26.30 | 26.01 | 26.77 | 26.36 | 69.8 | 85.1 | 89 | 92 | 102 | 94.20 | 153.6 |
| 2 | 19.60 | 19.20 | 19.44 | 19.41 | 75.0 | 87.1 | 87 | 93 | 103 | 94.19 | 261.0 |
| 3 | 15.33 | 15.43 | 15.15 | 15.30 | 77.4 | 87.6 | 87 | 92 | 104 | 94.18 | 401.6 |
| 4 | 13.76 | 13.01 | 13.32 | 13.36 | 77.0 | 85.8 | 81 | 89 | 105 | 91.52 | 504.0 |
| 5 | 10.55 | 10.67 | 10.45 | 10.56 | 79.1 | 85.8 | 82 | 86 | 105 | 90.85 | 773.3 |
| 6 | 9.43 | 9.67 | 9.23 | 9.44 | 80.0 | 86.0 | 82 | 87 | 103 | 90.52 | 943.7 |
| 7 | 8.32 | 8.43 | 8.54 | 8.43 | 77.3 | 83.0 | 80 | 81 | 102 | 87.51 | 1159.2 |
| 8 | 7.63 | 7.45 | 7.55 | 7.54 | 79.0 | 83.9 | 80 | 82 | 103 | 88.18 | 1423.3 |

### For 0.07% of $Al_2O_3$ Nanofluid

| Sr. No. | $t_1$ (Sec) | $t_2$ (Sec) | $t_3$ (Sec) | $t_{avg}$ (Sec) | $T_{in}$ (°C) | $T_{out}$ (°C) | $T_1$ (°C) | $T_2$ (°C) | $T_3$ (°C) | $T_s$ (°C) | $\Delta P$ (Pa) |
|---|---|---|---|---|---|---|---|---|---|---|---|
| 1 | 29.37 | 27.89 | 27.26 | 28.17 | 64.1 | 89.6 | 91 | 95 | 105 | 97.00 | 166.5 |
| 2 | 21.54 | 22.28 | 21.75 | 21.86 | 74.6 | 90.5 | 90 | 98 | 102 | 96.67 | 255.4 |
| 3 | 17.90 | 15.91 | 18.30 | 17.37 | 76.2 | 90.7 | 90 | 97 | 107 | 98.00 | 387.3 |
| 4 | 14.38 | 13.78 | 13.67 | 13.94 | 78.2 | 89.2 | 82 | 88 | 107 | 92.33 | 575.4 |
| 5 | 11.93 | 10.47 | 12.17 | 11.52 | 78.6 | 88.2 | 86 | 88 | 106 | 93.33 | 806.9 |
| 6 | 10.31 | 9.92 | 9.80 | 10.01 | 79.0 | 86.5 | 84 | 91 | 105 | 93.33 | 1047.5 |
| 7 | 9.50 | 8.62 | 8.45 | 8.86 | 78.5 | 85.0 | 82 | 82 | 104 | 89.33 | 1307.6 |
| 8 | 7.77 | 7.92 | 7.60 | 7.76 | 78.2 | 84.3 | 82 | 83 | 104 | 89.67 | 1677.7 |

### For 0.14% of $Al_2O_3$ Nanofluid

| Sr. No. | $t_1$ (Sec) | $t_2$ (Sec) | $t_3$ (Sec) | $t_{avg}$ (Sec) | $T_{in}$ (°C) | $T_{out}$ (°C) | $T_1$ (°C) | $T_2$ (°C) | $T_3$ (°C) | $T_s$ (°C) | $\Delta P$ (Pa) |
|---|---|---|---|---|---|---|---|---|---|---|---|
| 1 | 31.27 | 29.87 | 29.27 | 30.14 | 76.7 | 94.8 | 92 | 96 | 105 | 97.67 | 174.5 |
| 2 | 22.33 | 21.81 | 22.83 | 22.32 | 74.3 | 93.1 | 90 | 97 | 103 | 96.67 | 293.8 |
| 3 | 19.01 | 17.22 | 19.22 | 18.48 | 84.0 | 93.2 | 91 | 98 | 106 | 98.33 | 411.0 |
| 4 | 14.53 | 14.45 | 13.32 | 14.10 | 80.0 | 90.9 | 82 | 90 | 108 | 93.33 | 677.5 |
| 5 | 11.85 | 10.39 | 11.87 | 11.37 | 82.4 | 89.5 | 87 | 88 | 107 | 94.00 | 999.2 |
| 6 | 10.41 | 9.98 | 9.61 | 10.00 | 79.0 | 87.4 | 84 | 92 | 107 | 94.33 | 1265.7 |
| 7 | 9.54 | 8.65 | 9.65 | 9.28 | 79.9 | 85.8 | 81 | 83 | 106 | 90.00 | 1439.4 |
| 8 | 8.28 | 7.59 | 7.83 | 7.90 | 79.8 | 85.0 | 83 | 83 | 105 | 90.33 | 1957.5 |

### For 0.21% of $Al_2O_3$ Nanofluid

| Sr. No. | $t_1$ (Sec) | $t_2$ (Sec) | $t_3$ (Sec) | $t_{avg}$ (Sec) | $T_{in}$ (°C) | $T_{out}$ (°C) | $T_1$ (°C) | $T_2$ (°C) | $T_3$ (°C) | $T_s$ (°C) | $\Delta P$ (Pa) |
|---|---|---|---|---|---|---|---|---|---|---|---|
| 1 | 32.31 | 31.06 | 30.21 | 31.19 | 82.3 | 98.9 | 92 | 97 | 105 | 98.00 | 190.2 |
| 2 | 22.36 | 23.00 | 22.78 | 22.71 | 80.0 | 95.9 | 91 | 97 | 103 | 97.00 | 332.5 |
| 3 | 18.89 | 17.21 | 19.44 | 18.51 | 85.5 | 95.7 | 90 | 99 | 108 | 99.00 | 480.1 |
| 4 | 14.50 | 14.02 | 13.64 | 14.05 | 83.7 | 93.2 | 83 | 90 | 110 | 94.33 | 800.5 |
| 5 | 12.71 | 11.49 | 12.79 | 12.33 | 84.6 | 92.4 | 88 | 90 | 107 | 95.00 | 998.4 |
| 6 | 10.87 | 10.49 | 10.07 | 10.48 | 86.5 | 89.9 | 86 | 91 | 109 | 95.33 | 1357.2 |
| 7 | 9.50 | 8.81 | 9.89 | 9.40 | 81.9 | 87.8 | 83 | 84 | 109 | 92.00 | 1652.5 |
| 8 | 8.52 | 7.85 | 8.45 | 8.27 | 81.4 | 86.7 | 85 | 83 | 107 | 91.67 | 2105.0 |

# ANNEXURE II

## Estimated Parameter of Plain Tube
### For Water

| Sr. No. | m (kg/sec) | $\Delta T$ ($^0$C) | $T_b$ ($^0$C) | Q (W) | h (W/m²K) | Re | Nu | Pr | $\frac{N_{ut}}{N_{up}}$ | f | $\frac{f_t}{f_p}$ | η |
|---|---|---|---|---|---|---|---|---|---|---|---|---|
| 1 | 2.5 | 12 | 61.0 | 2057 | 975 | 5665 | 30 | 2.9 | 1.00 | 0.056 | 1.00 | 1.00 |
| 2 | 3.4 | 10.3 | 65.7 | 2372 | 1370 | 8205 | 42 | 2.7 | 1.00 | 0.049 | 1.00 | 1.00 |
| 3 | 4.0 | 9.1 | 67.9 | 2516 | 1644 | 10128 | 50 | 2.6 | 1.00 | 0.046 | 1.00 | 1.00 |
| 4 | 5.0 | 7.5 | 69.4 | 2560 | 1955 | 12668 | 59 | 2.6 | 1.00 | 0.042 | 1.00 | 1.00 |
| 5 | 6.1 | 6.3 | 71.2 | 2613 | 2305 | 15795 | 69 | 2.5 | 1.00 | 0.040 | 1.00 | 1.00 |
| 6 | 6.9 | 5.3 | 73.0 | 2482 | 2591 | 17160 | 78 | 2.6 | 1.00 | 0.037 | 1.00 | 1.00 |
| 7 | 7.8 | 4.3 | 74.9 | 2282 | 2861 | 20750 | 86 | 2.4 | 1.00 | 0.035 | 1.00 | 1.00 |
| 8 | 8.7 | 3.3 | 76.7 | 1950 | 3131 | 23097 | 94 | 2.4 | 1.00 | 0.033 | 1.00 | 1.00 |

### For 0.07% of Al$_2$O$_3$ Nanofluid

| Sr. No. | m (kg/sec) | $\Delta T$ ($^0$C) | $T_b$ ($^0$C) | Q (W) | h (W/m²K) | Re | Nu | Pr | $\frac{N_{ut}}{N_{up}}$ | f | $\frac{f_t}{f_p}$ | η |
|---|---|---|---|---|---|---|---|---|---|---|---|---|
| 1 | 2.1 | 14.4 | 73.1 | 2062 | 1418 | 5861 | 35 | 1.9 | 1.00 | 0.057 | 1.00 | 1.00 |
| 2 | 2.8 | 9.4 | 79.2 | 1801 | 1791 | 8425 | 44 | 1.8 | 1.00 | 0.050 | 1.00 | 1.00 |
| 3 | 3.6 | 8.2 | 81.0 | 1989 | 2279 | 10934 | 56 | 1.7 | 1.00 | 0.047 | 1.00 | 1.00 |
| 4 | 4.3 | 6.7 | 81.5 | 1933 | 2615 | 13206 | 64 | 1.7 | 1.00 | 0.043 | 1.00 | 1.00 |
| 5 | 5.4 | 5.5 | 81.7 | 1973 | 3090 | 16498 | 76 | 1.7 | 1.00 | 0.041 | 1.00 | 1.00 |
| 6 | 6.0 | 4.9 | 80.8 | 1979 | 3474 | 18299 | 85 | 1.7 | 1.00 | 0.038 | 1.00 | 1.00 |
| 7 | 7.0 | 4.3 | 79.8 | 1998 | 3791 | 20868 | 93 | 1.8 | 1.00 | 0.036 | 1.00 | 1.00 |
| 8 | 7.7 | 3.7 | 80.3 | 1916 | 4224 | 23217 | 104 | 1.8 | 1.00 | 0.034 | 1.00 | 1.00 |

### For 0.14% of Al$_2$O$_3$ Nanofluid

| Sr. No. | m (kg/sec) | $\Delta T$ ($^0$C) | $T_b$ ($^0$C) | Q (W) | h (W/m²K) | Re | Nu | Pr | $\frac{N_{ut}}{N_{up}}$ | f | $\frac{f_t}{f_p}$ | η |
|---|---|---|---|---|---|---|---|---|---|---|---|---|
| 1 | 2.0 | 18.8 | 73.3 | 2509 | 1914 | 5692 | 39 | 1.5 | 1.00 | 0.058 | 1.00 | 1.00 |
| 2 | 2.8 | 13.1 | 76.5 | 2416 | 2297 | 8189 | 47 | 1.5 | 1.00 | 0.051 | 1.00 | 1.00 |
| 3 | 3.3 | 11.1 | 79.2 | 2469 | 2883 | 10141 | 59 | 1.4 | 1.00 | 0.049 | 1.00 | 1.00 |
| 4 | 4.3 | 8.7 | 79.1 | 2482 | 3254 | 12945 | 67 | 1.4 | 1.00 | 0.044 | 1.00 | 1.00 |
| 5 | 5.4 | 7.3 | 79.2 | 2644 | 3932 | 16498 | 80 | 1.4 | 1.00 | 0.042 | 1.00 | 1.00 |
| 6 | 6.0 | 6.2 | 78.5 | 2484 | 4200 | 18250 | 86 | 1.4 | 1.00 | 0.039 | 1.00 | 1.00 |
| 7 | 6.7 | 5.6 | 77.4 | 2497 | 4518 | 19952 | 93 | 1.4 | 1.00 | 0.037 | 1.00 | 1.00 |
| 8 | 7.6 | 5.2 | 77.2 | 2657 | 5097 | 22811 | 104 | 1.4 | 1.00 | 0.035 | 1.00 | 1.00 |

### For 0.21% of Al$_2$O$_3$ Nanofluid

| Sr No | m (kg/sec) | $\Delta t$ (K) | $T_b$ (K) | Q (W) | h (W/m²K) | Re | Nu | Pr | Nut/Nup | f | ft/fp | η |
|---|---|---|---|---|---|---|---|---|---|---|---|---|
| 1 | 1.9 | 18.9 | 80.3 | 2402 | 2559 | 6030 | 44 | 1.1 | 1.00 | 0.060 | 1.00 | 1.00 |
| 2 | 2.7 | 15.8 | 79.7 | 2822 | 3248 | 8602 | 55 | 1.1 | 1.00 | 0.054 | 1.00 | 1.00 |
| 3 | 3.3 | 10.8 | 83.7 | 2376 | 3722 | 10940 | 63 | 1.1 | 1.00 | 0.049 | 1.00 | 1.00 |
| 4 | 4.3 | 9.4 | 82.2 | 2658 | 4391 | 13862 | 75 | 1.1 | 1.00 | 0.046 | 1.00 | 1.00 |
| 5 | 5.0 | 7.1 | 83.6 | 2325 | 4882 | 16426 | 83 | 1.1 | 1.00 | 0.043 | 1.00 | 1.00 |
| 6 | 5.7 | 6.5 | 81.5 | 2442 | 5301 | 18819 | 90 | 1.1 | 1.00 | 0.041 | 1.00 | 1.00 |
| 7 | 6.6 | 5.8 | 80.2 | 2530 | 5762 | 20558 | 98 | 1.1 | 1.00 | 0.039 | 1.00 | 1.00 |
| 8 | 7.3 | 5.3 | 79.7 | 2544 | 6198 | 22723 | 106 | 1.1 | 1.00 | 0.036 | 1.00 | 1.00 |

## Estimated Parameter of Single Twisted Tape
### For Water

| Sr. No. | m (kg/sec) | ΔT (°C) | $T_b$ (°C) | Q (W) | h (W/m²K) | Re | Nu | Pr | $\frac{N_{ut}}{N_{up}}$ | f | $\frac{f_t}{f_p}$ | η |
|---|---|---|---|---|---|---|---|---|---|---|---|---|
| 1 | 2.4 | 13.6 | 69.8 | 2212 | 1300 | 6114 | 39 | 2.6 | 1.32 | 0.102 | 1.83 | 0.91 |
| 2 | 3.2 | 10.8 | 75.0 | 2353 | 1766 | 8751 | 53 | 2.4 | 1.27 | 0.089 | 1.83 | 0.88 |
| 3 | 3.8 | 9 | 77.8 | 2336 | 2057 | 10810 | 62 | 2.3 | 1.24 | 0.086 | 1.86 | 0.85 |
| 4 | 4.8 | 7.5 | 77.9 | 2437 | 2421 | 13535 | 72 | 2.3 | 1.23 | 0.082 | 1.95 | 0.83 |
| 5 | 5.8 | 6.2 | 79.3 | 2454 | 2878 | 16689 | 86 | 2.3 | 1.21 | 0.078 | 1.97 | 0.84 |
| 6 | 6.5 | 5.4 | 79.0 | 2406 | 3138 | 18557 | 94 | 2.3 | 1.20 | 0.077 | 2.08 | 0.80 |
| 7 | 7.4 | 5.1 | 77.5 | 2569 | 3481 | 20976 | 104 | 2.3 | 1.21 | 0.075 | 2.13 | 0.80 |
| 8 | 8.2 | 4.7 | 77.5 | 2639 | 3793 | 23102 | 114 | 2.3 | 1.21 | 0.074 | 2.25 | 0.78 |

### For 0.07% of $Al_2O_3$ Nanofluid

| Sr. No. | m (kg/sec) | ΔT (°C) | $T_b$ (°C) | Q (W) | h (W/m²K) | Re | Nu | Pr | $\frac{N_{ut}}{N_{up}}$ | f | $\frac{f_t}{f_p}$ | η |
|---|---|---|---|---|---|---|---|---|---|---|---|---|
| 1 | 2.1 | 16.7 | 73.1 | 2400 | 1653 | 6030 | 41 | 1.9 | 1.34 | 0.105 | 1.89 | 1.04 |
| 2 | 2.8 | 12.63 | 78.1 | 2399 | 2236 | 8476 | 55 | 1.8 | 1.31 | 0.092 | 1.89 | 0.97 |
| 3 | 3.6 | 10.03 | 80.2 | 2399 | 2612 | 10922 | 64 | 1.7 | 1.26 | 0.089 | 1.92 | 0.88 |
| 4 | 4.3 | 8.279 | 80.8 | 2399 | 3085 | 13234 | 76 | 1.7 | 1.24 | 0.084 | 1.99 | 0.89 |
| 5 | 5.4 | 6.513 | 81.6 | 2348 | 3623 | 16468 | 89 | 1.7 | 1.23 | 0.081 | 2.03 | 0.86 |
| 6 | 6.0 | 6.16 | 80.3 | 2486 | 4129 | 18435 | 101 | 1.7 | 1.25 | 0.080 | 2.12 | 0.85 |
| 7 | 7.0 | 5.048 | 79.7 | 2363 | 4436 | 20884 | 109 | 1.8 | 1.23 | 0.078 | 2.22 | 0.83 |
| 8 | 7.7 | 4.605 | 79.8 | 2398 | 4907 | 23227 | 121 | 1.8 | 1.22 | 0.078 | 2.34 | 0.84 |

### For 0.14% of $Al_2O_3$ Nanofluid

| Sr. No. | m (kg/sec) | ΔT (°C) | $T_b$ (°C) | Q (W) | h (W/m²K) | Re | Nu | Pr | $\frac{N_{ut}}{N_{up}}$ | f | $\frac{f_t}{f_p}$ | η |
|---|---|---|---|---|---|---|---|---|---|---|---|---|
| 1 | 2.0 | 17.86 | 73.0 | 2400 | 1718 | 5489 | 43 | 1.9 | 1.41 | 0.109 | 1.92 | 1.14 |
| 2 | 2.8 | 13.22 | 77.4 | 2460 | 2327 | 7999 | 57 | 1.8 | 1.34 | 0.099 | 1.92 | 1.09 |
| 3 | 3.3 | 11.21 | 79.1 | 2520 | 2722 | 9904 | 67 | 1.8 | 1.31 | 0.095 | 1.95 | 0.99 |
| 4 | 4.3 | 12.11 | 78.1 | 3438 | 3884 | 13108 | 79 | 1.4 | 1.25 | 0.088 | 2.04 | 0.95 |
| 5 | 5.4 | 9.788 | 78.9 | 3541 | 4731 | 16705 | 97 | 1.4 | 1.26 | 0.085 | 2.06 | 0.89 |
| 6 | 6.0 | 6.377 | 79.2 | 2580 | 4243 | 17823 | 104 | 1.8 | 1.25 | 0.083 | 2.17 | 0.88 |
| 7 | 6.7 | 5.828 | 78.4 | 2610 | 4749 | 19487 | 117 | 1.8 | 1.26 | 0.082 | 2.23 | 0.87 |
| 8 | 7.6 | 5.156 | 78.5 | 2640 | 5217 | 22279 | 129 | 1.8 | 1.23 | 0.080 | 2.35 | 0.86 |

### For 0.21% of $Al_2O_3$ Nanofluid

| Sr. No. | m (kg/sec) | ΔT (°C) | $T_b$ (°C) | Q (W) | h (W/m²K) | Re | Nu | Pr | $\frac{N_{ut}}{N_{up}}$ | f | $\frac{f_t}{f_p}$ | η |
|---|---|---|---|---|---|---|---|---|---|---|---|---|
| 1 | 1.9 | 23.4 | 78.7 | 2968 | 2703 | 5596 | 46 | 1.2 | 1.44 | 0.113 | 1.97 | 1.25 |
| 2 | 2.7 | 17 | 81.5 | 3041 | 3721 | 8294 | 64 | 1.2 | 1.38 | 0.105 | 1.97 | 1.13 |
| 3 | 3.3 | 14.75 | 83.4 | 3240 | 4504 | 10560 | 77 | 1.1 | 1.33 | 0.096 | 2.05 | 1.09 |
| 4 | 4.3 | 9.062 | 84.5 | 2552 | 4884 | 13056 | 84 | 1.2 | 1.32 | 0.093 | 2.10 | 0.99 |
| 5 | 5.0 | 7.739 | 85.4 | 2552 | 5991 | 15673 | 102 | 1.1 | 1.29 | 0.088 | 2.19 | 0.96 |
| 6 | 5.7 | 6.757 | 83.6 | 2553 | 6517 | 18164 | 111 | 1.1 | 1.27 | 0.088 | 2.23 | 0.91 |
| 7 | 6.6 | 5.902 | 82.7 | 2553 | 7352 | 20558 | 126 | 1.1 | 1.28 | 0.084 | 2.26 | 0.88 |
| 8 | 7.3 | 5.341 | 82.2 | 2554 | 8135 | 22723 | 139 | 1.1 | 1.25 | 0.083 | 2.42 | 0.87 |

## Estimated Parameter of Dual Co-Tapes
### For Water

| Sr. No. | m (kg/sec) | $\Delta T$ ($^0$C) | $T_b$ ($^0$C) | Q (W) | h (W/m²K) | Re | Nu | Pr | $\dfrac{N_{ut}}{N_{up}}$ | f | $\dfrac{f_t}{f_p}$ | η |
|---|---|---|---|---|---|---|---|---|---|---|---|---|
| 1 | 2.1 | 15.4 | 67.7 | 2260 | 1247 | 5378 | 38 | 2.6 | 1.27 | 0.196 | 3.58 | 0.83 |
| 2 | 2.8 | 11.6 | 74.8 | 2216 | 1647 | 7674 | 49 | 2.4 | 1.20 | 0.188 | 3.85 | 0.75 |
| 3 | 3.5 | 9.6 | 78.4 | 2307 | 2024 | 10135 | 60 | 2.3 | 1.17 | 0.175 | 3.84 | 0.76 |
| 4 | 4.3 | 7.9 | 77.8 | 2333 | 2351 | 12298 | 70 | 2.3 | 1.14 | 0.175 | 4.14 | 0.74 |
| 5 | 5.3 | 6.4 | 78.8 | 2319 | 2823 | 15278 | 84 | 2.3 | 1.12 | 0.165 | 4.22 | 0.74 |
| 6 | 6.0 | 5.7 | 79.7 | 2344 | 3137 | 17339 | 94 | 2.3 | 1.11 | 0.168 | 4.54 | 0.74 |
| 7 | 6.9 | 5.1 | 78.9 | 2400 | 3467 | 19848 | 104 | 2.3 | 1.11 | 0.158 | 4.49 | 0.73 |
| 8 | 7.8 | 4.6 | 77.6 | 2439 | 3781 | 22085 | 113 | 2.3 | 1.11 | 0.155 | 4.63 | 0.74 |

### For 0.07% of $Al_2O_3$ Nanofluid

| Sr. No. | m (kg/sec) | $\Delta T$ ($^0$C) | $T_b$ ($^0$C) | Q (W) | h (W/m²K) | Re | Nu | Pr | $\dfrac{N_{ut}}{N_{up}}$ | f | $\dfrac{f_t}{f_p}$ | η |
|---|---|---|---|---|---|---|---|---|---|---|---|---|
| 1 | 2.0 | 15.0 | 76.5 | 2031 | 1640 | 5810 | 40 | 1.8 | 1.36 | 0.210 | 3.81 | 0.97 |
| 2 | 2.7 | 11.8 | 80.5 | 2177 | 2358 | 8337 | 58 | 1.7 | 1.34 | 0.192 | 4.02 | 0.90 |
| 3 | 3.3 | 9.6 | 82.0 | 2132 | 2634 | 10190 | 65 | 1.7 | 1.27 | 0.185 | 4.09 | 0.83 |
| 4 | 4.3 | 8.0 | 82.2 | 2304 | 3334 | 13180 | 82 | 1.7 | 1.34 | 0.176 | 4.29 | 0.88 |
| 5 | 5.4 | 6.3 | 82.5 | 2285 | 3864 | 16765 | 95 | 1.7 | 1.31 | 0.172 | 4.33 | 0.82 |
| 6 | 6.1 | 5.7 | 81.3 | 2302 | 4282 | 18581 | 105 | 1.7 | 1.32 | 0.165 | 4.57 | 0.82 |
| 7 | 6.6 | 4.8 | 80.4 | 2129 | 4330 | 20068 | 106 | 1.7 | 1.23 | 0.163 | 4.70 | 0.75 |
| 8 | 7.7 | 4.5 | 80.2 | 2324 | 5010 | 22987 | 123 | 1.8 | 1.28 | 0.159 | 4.97 | 0.77 |

### For 0.14% of $Al_2O_3$ Nanofluid

| Sr. No. | m (kg/sec) | $\Delta T$ ($^0$C) | $T_b$ ($^0$C) | Q (W) | h (W/m²K) | Re | Nu | Pr | $\dfrac{N_{ut}}{N_{up}}$ | f | $\dfrac{f_t}{f_p}$ | η |
|---|---|---|---|---|---|---|---|---|---|---|---|---|
| 1 | 2.0 | 19.0 | 78.5 | 2523 | 2405 | 6055 | 49 | 1.4 | 1.50 | 0.220 | 3.93 | 1.12 |
| 2 | 2.8 | 13.4 | 81.5 | 2461 | 3094 | 8797 | 63 | 1.4 | 1.45 | 0.205 | 4.14 | 1.03 |
| 3 | 3.3 | 11.3 | 82.6 | 2521 | 3574 | 10757 | 73 | 1.3 | 1.38 | 0.188 | 4.22 | 0.94 |
| 4 | 4.3 | 9.0 | 82.5 | 2551 | 4205 | 13570 | 86 | 1.4 | 1.35 | 0.183 | 4.43 | 0.95 |
| 5 | 5.4 | 7.3 | 82.3 | 2640 | 4939 | 17294 | 101 | 1.4 | 1.31 | 0.173 | 4.48 | 0.88 |
| 6 | 6.0 | 7.2 | 80.7 | 2868 | 5596 | 19359 | 114 | 1.3 | 1.33 | 0.170 | 4.73 | 0.88 |
| 7 | 6.7 | 5.9 | 79.8 | 2612 | 5611 | 20454 | 115 | 1.4 | 1.31 | 0.166 | 4.87 | 0.85 |
| 8 | 7.6 | 5.3 | 79.7 | 2703 | 6280 | 23384 | 128 | 1.4 | 1.30 | 0.167 | 5.15 | 0.81 |

### For 0.21% of $Al_2O_3$ Nanofluid

| Sr. No. | m (kg/sec) | $\Delta T$ ($^0$C) | $T_b$ ($^0$C) | Q (W) | h (W/m²K) | Re | Nu | Pr | $\dfrac{N_{ut}}{N_{up}}$ | f | $\dfrac{f_t}{f_p}$ | η |
|---|---|---|---|---|---|---|---|---|---|---|---|---|
| 1 | 1.9 | 20.1 | 83.5 | 2567 | 3219 | 6213 | 55 | 1.1 | 1.54 | 0.230 | 4.05 | 1.20 |
| 2 | 2.7 | 14.0 | 84.4 | 2488 | 3906 | 9064 | 66 | 1.1 | 1.44 | 0.206 | 4.27 | 1.05 |
| 3 | 3.3 | 12.4 | 85.7 | 2707 | 4713 | 11257 | 80 | 1.0 | 1.39 | 0.193 | 4.35 | 1.02 |
| 4 | 4.3 | 9.1 | 85.4 | 2568 | 5446 | 14460 | 92 | 1.1 | 1.36 | 0.186 | 4.58 | 0.98 |
| 5 | 5.0 | 7.7 | 85.7 | 2534 | 6250 | 16707 | 106 | 1.1 | 1.34 | 0.177 | 4.63 | 0.93 |
| 6 | 5.8 | 6.7 | 84.1 | 2537 | 6962 | 19621 | 118 | 1.0 | 1.35 | 0.178 | 4.90 | 0.90 |
| 7 | 6.5 | 5.9 | 82.4 | 2549 | 7045 | 21151 | 120 | 1.1 | 1.32 | 0.170 | 5.04 | 0.88 |
| 8 | 7.3 | 5.3 | 82.0 | 2569 | 7852 | 23416 | 134 | 1.1 | 1.33 | 0.168 | 5.34 | 0.84 |

## Estimated Parameter of Dual C-Tapes

### For Water

| Sr. No. | m (kg/sec) | $\Delta T$ ($^0C$) | $T_b$ ($^0C$) | Q (W) | h (W/m²K) | Re | Nu | Pr | $\dfrac{N_{ut}}{N_{up}}$ | f | $\dfrac{f_t}{f_p}$ | η |
|---|---|---|---|---|---|---|---|---|---|---|---|---|
| 1 | 2.3 | 14.9 | 70.7 | 2366 | 1473 | 6048 | 44 | 2.5 | 1.51 | 0.219 | 3.94 | 0.81 |
| 2 | 3.2 | 11.1 | 76.5 | 2430 | 1991 | 8898 | 60 | 2.3 | 1.42 | 0.202 | 4.15 | 0.75 |
| 3 | 4.0 | 9 | 79.6 | 2450 | 2348 | 11477 | 70 | 2.3 | 1.40 | 0.194 | 4.22 | 0.74 |
| 4 | 4.7 | 7.9 | 79.0 | 2507 | 2737 | 13384 | 82 | 2.3 | 1.37 | 0.185 | 4.42 | 0.71 |
| 5 | 5.7 | 6.4 | 78.6 | 2496 | 3239 | 16444 | 97 | 2.3 | 1.38 | 0.177 | 4.46 | 0.71 |
| 6 | 6.4 | 5.7 | 80.2 | 2473 | 3563 | 18297 | 106 | 2.3 | 1.37 | 0.173 | 4.70 | 0.70 |
| 7 | 7.3 | 5.3 | 79.1 | 2626 | 3991 | 20896 | 119 | 2.3 | 1.39 | 0.169 | 4.82 | 0.70 |
| 8 | 8.1 | 4.8 | 78.0 | 2659 | 4294 | 23073 | 128 | 2.3 | 1.39 | 0.167 | 5.09 | 0.69 |

### For 0.07% of $Al_2O_3$ Nanofluid

| Sr. No. | m (kg/sec) | $\Delta T$ ($^0C$) | $T_b$ ($^0C$) | Q (W) | h (W/m²K) | Re | Nu | Pr | $\dfrac{N_{ut}}{N_{up}}$ | f | $\dfrac{f_t}{f_p}$ | η |
|---|---|---|---|---|---|---|---|---|---|---|---|---|
| 1 | 2.2 | 17.3 | 76.7 | 2526 | 2064 | 6109 | 51 | 1.9 | 1.60 | 0.223 | 4.10 | 1.19 |
| 2 | 2.8 | 12.5 | 80.3 | 2344 | 2494 | 8364 | 61 | 1.8 | 1.46 | 0.211 | 4.24 | 0.93 |
| 3 | 3.5 | 10.2 | 82.1 | 2414 | 3014 | 10766 | 74 | 1.7 | 1.45 | 0.198 | 4.44 | 0.93 |
| 4 | 4.4 | 8.2 | 82.3 | 2432 | 3554 | 13412 | 87 | 1.7 | 1.43 | 0.192 | 4.52 | 0.92 |
| 5 | 5.3 | 6.9 | 82.5 | 2472 | 4191 | 16228 | 103 | 1.7 | 1.38 | 0.184 | 4.71 | 0.87 |
| 6 | 6.1 | 5.9 | 81.4 | 2432 | 4593 | 18681 | 113 | 1.7 | 1.40 | 0.180 | 4.82 | 0.86 |
| 7 | 6.9 | 5.4 | 80.5 | 2503 | 5180 | 20638 | 127 | 1.8 | 1.38 | 0.170 | 5.10 | 0.88 |
| 8 | 7.8 | 4.6 | 80.5 | 2434 | 5503 | 23548 | 135 | 1.8 | 1.42 | 0.171 | 5.34 | 0.83 |

### For 0.14% of $Al_2O_3$ Nanofluid

| Sr. No. | m (kg/sec) | $\Delta T$ ($^0C$) | $T_b$ ($^0C$) | Q (W) | h (W/m²K) | Re | Nu | Pr | $\dfrac{N_{ut}}{N_{up}}$ | f | $\dfrac{f_t}{f_p}$ | η |
|---|---|---|---|---|---|---|---|---|---|---|---|---|
| 1 | 2.0 | 18.0 | 80.8 | 2432 | 2696 | 6135 | 55 | 1.4 | 1.61 | 0.228 | 4.22 | 1.23 |
| 2 | 2.7 | 13.3 | 82.2 | 2428 | 3223 | 8482 | 66 | 1.4 | 1.54 | 0.213 | 4.37 | 1.05 |
| 3 | 3.3 | 11.3 | 83.6 | 2486 | 3860 | 10363 | 79 | 1.4 | 1.47 | 0.209 | 4.54 | 0.99 |
| 4 | 4.3 | 8.9 | 83.1 | 2585 | 4514 | 13588 | 92 | 1.4 | 1.45 | 0.194 | 4.81 | 1.00 |
| 5 | 5.4 | 7.1 | 83.2 | 2544 | 5286 | 16848 | 108 | 1.4 | 1.41 | 0.186 | 4.82 | 0.92 |
| 6 | 6.1 | 7.6 | 81.0 | 2535 | 6311 | 19157 | 129 | 1.4 | 1.42 | 0.182 | 5.04 | 0.92 |
| 7 | 6.6 | 5.7 | 80.9 | 2520 | 6356 | 20174 | 130 | 1.4 | 1.40 | 0.184 | 5.23 | 0.87 |
| 8 | 7.7 | 5.8 | 79.9 | 2612 | 7281 | 23694 | 149 | 1.4 | 1.43 | 0.176 | 5.48 | 0.85 |

### For 0.21% of $Al_2O_3$ Nanofluid

| Sr. No. | m (kg/sec) | $\Delta T$ ($^0C$) | $T_b$ ($^0C$) | Q (W) | h (W/m²K) | Re | Nu | Pr | $\dfrac{N_{ut}}{N_{up}}$ | f | $\dfrac{f_t}{f_p}$ | η |
|---|---|---|---|---|---|---|---|---|---|---|---|---|
| 1 | 2.0 | 19.6 | 84.2 | 2522 | 3343 | 6404 | 57 | 1.1 | 1.69 | 0.237 | 4.31 | 1.25 |
| 2 | 2.7 | 14.6 | 85.0 | 2576 | 4305 | 8901 | 73 | 1.1 | 1.56 | 0.224 | 4.51 | 1.13 |
| 3 | 3.3 | 11.6 | 86.7 | 2515 | 4897 | 11046 | 83 | 1.1 | 1.50 | 0.207 | 4.73 | 1.03 |
| 4 | 4.3 | 9.4 | 85.8 | 2687 | 6055 | 14555 | 103 | 1.1 | 1.50 | 0.202 | 4.87 | 1.07 |
| 5 | 4.9 | 7.7 | 86.1 | 2516 | 6585 | 16587 | 112 | 1.1 | 1.47 | 0.191 | 4.93 | 0.96 |
| 6 | 5.8 | 6.5 | 84.7 | 2468 | 7645 | 19521 | 130 | 1.1 | 1.48 | 0.191 | 5.22 | 0.96 |
| 7 | 6.5 | 7.0 | 82.5 | 2984 | 8376 | 21511 | 142 | 1.1 | 1.45 | 0.183 | 5.47 | 0.92 |
| 8 | 7.4 | 5.3 | 82.5 | 2586 | 8837 | 23853 | 150 | 1.1 | 1.46 | 0.184 | 5.52 | 0.89 |

## Estimated Parameter of Triple Co-Tapes

### For Water

| Sr. No. | m (kg/sec) | $\Delta T$ ($^0C$) | $T_b$ ($^0C$) | Q (W) | h (W/m²K) | Re | Nu | Pr | $\dfrac{N_{ut}}{N_{up}}$ | f | $\dfrac{f_t}{f_p}$ | η |
|---|---|---|---|---|---|---|---|---|---|---|---|---|
| 1 | 2.3 | 15.4 | 73.7 | 2424 | 1689 | 6238 | 51 | 2.4 | 1.72 | 0.306 | 5.59 | 0.83 |
| 2 | 3.2 | 11.6 | 78.0 | 2490 | 2261 | 8942 | 68 | 2.3 | 1.62 | 0.282 | 5.78 | 0.77 |
| 3 | 4.0 | 9.1 | 80.8 | 2472 | 2661 | 11598 | 79 | 2.2 | 1.58 | 0.271 | 5.95 | 0.74 |
| 4 | 4.6 | 8.3 | 80.9 | 2574 | 3149 | 13398 | 94 | 2.2 | 1.58 | 0.259 | 6.14 | 0.73 |
| 5 | 5.6 | 6.7 | 80.3 | 2529 | 3677 | 16118 | 110 | 2.2 | 1.57 | 0.248 | 6.33 | 0.72 |
| 6 | 6.5 | 5.8 | 82.3 | 2547 | 4096 | 18749 | 122 | 2.2 | 1.58 | 0.242 | 6.53 | 0.72 |
| 7 | 7.0 | 5.6 | 79.8 | 2677 | 4539 | 20411 | 136 | 2.2 | 1.57 | 0.236 | 6.73 | 0.71 |
| 8 | 8.0 | 5 | 79.7 | 2724 | 4917 | 23263 | 147 | 2.2 | 1.59 | 0.233 | 6.94 | 0.71 |

### For 0.07% of $Al_2O_3$ Nanofluid

| Sr. No. | m (kg/sec) | $\Delta T$ ($^0C$) | $T_b$ ($^0C$) | Q (W) | h (W/m²K) | Re | Nu | Pr | $\dfrac{N_{ut}}{N_{up}}$ | f | $\dfrac{f_t}{f_p}$ | η |
|---|---|---|---|---|---|---|---|---|---|---|---|---|
| 1 | 2.1 | 19.3 | 76.6 | 2775 | 2258 | 6193 | 56 | 1.8 | 1.87 | 0.312 | 5.72 | 1.07 |
| 2 | 2.8 | 12.9 | 81.2 | 2392 | 2727 | 8471 | 67 | 1.7 | 1.67 | 0.297 | 6.03 | 0.91 |
| 3 | 3.5 | 10.8 | 82.6 | 2524 | 3284 | 10783 | 81 | 1.7 | 1.63 | 0.276 | 6.13 | 0.90 |
| 4 | 4.3 | 9.1 | 82.8 | 2653 | 4058 | 13433 | 100 | 1.7 | 1.63 | 0.270 | 6.37 | 0.94 |
| 5 | 5.2 | 7.3 | 83.0 | 2562 | 4581 | 16254 | 112 | 1.7 | 1.56 | 0.254 | 6.50 | 0.85 |
| 6 | 6.0 | 6.8 | 81.7 | 2748 | 5362 | 18712 | 132 | 1.7 | 1.62 | 0.248 | 6.79 | 0.86 |
| 7 | 6.8 | 5.6 | 81.0 | 2543 | 5608 | 20665 | 138 | 1.7 | 1.55 | 0.247 | 7.11 | 0.82 |
| 8 | 7.8 | 5.2 | 80.7 | 2712 | 6377 | 23579 | 157 | 1.7 | 1.58 | 0.239 | 7.52 | 0.86 |

### For 0.14% of $Al_2O_3$ Nanofluid

| Sr. No. | m (kg/sec) | $\Delta T$ ($^0C$) | $T_b$ ($^0C$) | Q (W) | h (W/m²K) | Re | Nu | Pr | $\dfrac{N_{ut}}{N_{up}}$ | f | $\dfrac{f_t}{f_p}$ | η |
|---|---|---|---|---|---|---|---|---|---|---|---|---|
| 1 | 2.0 | 18.1 | 80.6 | 2409 | 2821 | 5711 | 58 | 1.5 | 2.06 | 0.324 | 5.78 | 1.15 |
| 2 | 2.7 | 15.6 | 80.8 | 2809 | 3613 | 8001 | 74 | 1.5 | 1.89 | 0.303 | 6.10 | 1.05 |
| 3 | 3.3 | 10.9 | 83.5 | 2376 | 4023 | 9903 | 82 | 1.4 | 1.77 | 0.294 | 6.33 | 0.93 |
| 4 | 4.3 | 9.4 | 82.7 | 2686 | 5048 | 12985 | 103 | 1.4 | 1.74 | 0.273 | 6.58 | 1.00 |
| 5 | 5.3 | 7.7 | 82.4 | 2724 | 5786 | 16100 | 118 | 1.4 | 1.62 | 0.262 | 6.72 | 0.90 |
| 6 | 6.0 | 6.5 | 81.5 | 2592 | 6475 | 18306 | 132 | 1.4 | 1.65 | 0.262 | 7.03 | 0.89 |
| 7 | 6.5 | 5.9 | 80.1 | 2553 | 6694 | 19484 | 137 | 1.4 | 1.59 | 0.251 | 7.37 | 0.85 |
| 8 | 7.6 | 5.2 | 79.9 | 2653 | 7578 | 22884 | 155 | 1.4 | 1.64 | 0.253 | 7.57 | 0.87 |

### For 0.21% of $Al_2O_3$ Nanofluid

| Sr. No. | m (kg/sec) | $\Delta T$ ($^0C$) | $T_b$ ($^0C$) | Q (W) | h (W/m²K) | Re | Nu | Pr | $\dfrac{N_{ut}}{N_{up}}$ | f | $\dfrac{f_t}{f_p}$ | η |
|---|---|---|---|---|---|---|---|---|---|---|---|---|
| 1 | 1.9 | 20.1 | 84.4 | 2559 | 3763 | 6050 | 64 | 1.1 | 2.21 | 0.324 | 5.95 | 1.26 |
| 2 | 2.7 | 14.3 | 85.2 | 2492 | 4764 | 8404 | 81 | 1.1 | 1.91 | 0.316 | 6.41 | 1.11 |
| 3 | 3.3 | 11.6 | 86.3 | 2491 | 5243 | 10682 | 89 | 1.1 | 1.83 | 0.292 | 6.47 | 0.99 |
| 4 | 4.3 | 9.1 | 85.6 | 2559 | 6518 | 13908 | 111 | 1.1 | 1.75 | 0.282 | 6.80 | 1.02 |
| 5 | 4.9 | 7.7 | 85.8 | 2491 | 7402 | 16041 | 126 | 1.1 | 1.67 | 0.279 | 6.95 | 0.96 |
| 6 | 5.8 | 6.8 | 83.9 | 2561 | 8206 | 18878 | 140 | 1.1 | 1.70 | 0.266 | 7.28 | 0.92 |
| 7 | 6.4 | 5.9 | 82.5 | 2492 | 8405 | 20078 | 143 | 1.1 | 1.60 | 0.255 | 7.64 | 0.86 |
| 8 | 7.3 | 5.3 | 81.9 | 2561 | 9431 | 22806 | 161 | 1.1 | 1.67 | 0.257 | 8.17 | 0.88 |

## Estimated Parameter of Triple Counter Tapes

### For Water

| Sr. No. | m (kg/sec) | ΔT (°C) | $T_b$ (°C) | Q (W) | h (W/m²K) | Re | Nu | Pr | $N_{ut}/N_{up}$ | f | $f_t/f_p$ | η |
|---|---|---|---|---|---|---|---|---|---|---|---|---|
| 1 | 2.3 | 15.6 | 74.1 | 2415 | 1766 | 6135 | 53 | 2.4 | 1.72 | 0.321 | 5.87 | 0.85 |
| 2 | 3.1 | 12 | 79.0 | 2535 | 2392 | 8907 | 71 | 2.3 | 1.62 | 0.297 | 6.07 | 0.80 |
| 3 | 4.0 | 9.3 | 81.8 | 2512 | 2833 | 11670 | 85 | 2.2 | 1.58 | 0.284 | 6.25 | 0.78 |
| 4 | 4.5 | 8.6 | 80.9 | 2636 | 3324 | 13243 | 99 | 2.2 | 1.58 | 0.272 | 6.44 | 0.76 |
| 5 | 5.7 | 6.8 | 80.6 | 2627 | 3948 | 16689 | 118 | 2.2 | 1.57 | 0.260 | 6.65 | 0.76 |
| 6 | 6.3 | 6.2 | 81.5 | 2667 | 4382 | 18584 | 131 | 2.2 | 1.58 | 0.254 | 6.85 | 0.75 |
| 7 | 7.2 | 5.7 | 79.7 | 2796 | 4838 | 20942 | 144 | 2.2 | 1.57 | 0.248 | 7.07 | 0.74 |
| 8 | 8.0 | 5.1 | 79.8 | 2777 | 5243 | 23252 | 157 | 2.2 | 1.59 | 0.244 | 7.29 | 0.74 |

### For 0.07% of $Al_2O_3$ Nanofluid

| Sr. No. | m (kg/sec) | ΔT (°C) | $T_b$ (°C) | Q (W) | h (W/m²K) | Re | Nu | Pr | $N_{ut}/N_{up}$ | f | $f_t/f_p$ | η |
|---|---|---|---|---|---|---|---|---|---|---|---|---|
| 1 | 2.1 | 16.7 | 79.6 | 2397 | 2297 | 5853 | 57 | 1.9 | 1.91 | 0.334 | 6.01 | 1.08 |
| 2 | 2.8 | 13.1 | 82.6 | 2451 | 3085 | 8234 | 76 | 1.8 | 1.81 | 0.306 | 6.21 | 1.01 |
| 3 | 3.5 | 9.9 | 84.4 | 2331 | 3532 | 10618 | 87 | 1.7 | 1.70 | 0.296 | 6.56 | 0.96 |
| 4 | 4.3 | 7.9 | 84.4 | 2273 | 4122 | 13192 | 101 | 1.7 | 1.73 | 0.281 | 6.69 | 0.94 |
| 5 | 5.3 | 6.6 | 84.3 | 2357 | 4937 | 16191 | 121 | 1.7 | 1.68 | 0.274 | 6.83 | 0.93 |
| 6 | 6.0 | 5.9 | 83.0 | 2384 | 5524 | 18167 | 136 | 1.7 | 1.67 | 0.263 | 7.13 | 0.91 |
| 7 | 6.9 | 5.3 | 81.9 | 2457 | 6173 | 20562 | 152 | 1.8 | 1.66 | 0.260 | 7.47 | 0.89 |
| 8 | 7.7 | 4.6 | 81.8 | 2396 | 6598 | 23157 | 162 | 1.8 | 1.72 | 0.251 | 7.67 | 0.87 |

### For 0.14% of $Al_2O_3$ Nanofluid

| Sr. No. | m (kg/sec) | ΔT (°C) | $T_b$ (°C) | Q (W) | h (W/m²K) | Re | Nu | Pr | $N_{ut}/N_{up}$ | f | $f_t/f_p$ | η |
|---|---|---|---|---|---|---|---|---|---|---|---|---|
| 1 | 2.0 | 20.8 | 80.4 | 2751 | 3175 | 6181 | 65 | 1.4 | 2.15 | 0.347 | 6.19 | 1.27 |
| 2 | 2.7 | 12.5 | 84.2 | 2256 | 4000 | 8653 | 82 | 1.4 | 1.91 | 0.325 | 6.53 | 1.14 |
| 3 | 3.3 | 11.3 | 84.4 | 2480 | 4666 | 10581 | 95 | 1.3 | 1.85 | 0.306 | 6.64 | 1.05 |
| 4 | 4.3 | 9.1 | 83.5 | 2567 | 5339 | 13525 | 109 | 1.4 | 1.80 | 0.299 | 6.91 | 1.03 |
| 5 | 5.3 | 7.6 | 83.3 | 2690 | 6517 | 17006 | 133 | 1.4 | 1.73 | 0.284 | 7.06 | 1.00 |
| 6 | 6.0 | 7.1 | 81.7 | 2827 | 7382 | 19294 | 150 | 1.3 | 1.75 | 0.278 | 7.46 | 0.96 |
| 7 | 6.5 | 5.9 | 80.9 | 2571 | 7806 | 20350 | 159 | 1.4 | 1.72 | 0.272 | 7.59 | 0.93 |
| 8 | 7.6 | 5.7 | 80.2 | 2864 | 8647 | 23307 | 177 | 1.4 | 1.69 | 0.273 | 8.35 | 0.90 |

### For 0.21% of $Al_2O_3$ Nanofluid

| Sr. No. | m (kg/sec) | ΔT (°C) | $T_b$ (°C) | Q (W) | h (W/m²K) | Re | Nu | Pr | $N_{ut}/N_{up}$ | f | $f_t/f_p$ | η |
|---|---|---|---|---|---|---|---|---|---|---|---|---|
| 1 | 1.9 | 20.1 | 85.4 | 2542 | 4111 | 6376 | 70 | 1.1 | 2.29 | 0.365 | 6.25 | 1.35 |
| 2 | 2.7 | 14.3 | 85.7 | 2509 | 5121 | 8872 | 87 | 1.1 | 2.04 | 0.335 | 6.73 | 1.17 |
| 3 | 3.3 | 11.6 | 87.1 | 2509 | 5912 | 11146 | 100 | 1.0 | 1.90 | 0.328 | 6.86 | 1.09 |
| 4 | 4.3 | 10.4 | 85.2 | 2907 | 6946 | 14154 | 118 | 1.1 | 1.86 | 0.308 | 7.14 | 1.07 |
| 5 | 4.9 | 7.7 | 86.3 | 2510 | 8114 | 16542 | 138 | 1.1 | 1.79 | 0.302 | 7.30 | 1.03 |
| 6 | 5.7 | 6.8 | 84.4 | 2545 | 9101 | 19427 | 154 | 1.0 | 1.80 | 0.288 | 7.64 | 1.00 |
| 7 | 6.5 | 6.4 | 82.8 | 2707 | 9788 | 20942 | 167 | 1.1 | 1.69 | 0.293 | 8.02 | 0.98 |
| 8 | 7.2 | 5.3 | 82.3 | 2545 | 10501 | 23184 | 179 | 1.1 | 1.69 | 0.283 | 8.50 | 0.96 |

### Estimated Parameter of Quadruple Counter Tapes as Co-Quadruple Swirl Flow Generators

#### For Water

| Sr. No. | m (kg/sec) | ΔT (°C) | $T_b$ (°C) | Q (W) | h (W/m²K) | Re | Nu | Pr | $\frac{N_{ut}}{N_{up}}$ | f | $\frac{f_t}{f_p}$ | η |
|---|---|---|---|---|---|---|---|---|---|---|---|---|
| 1 | 2.3 | 15.3 | 76.6 | 2376 | 2102 | 6312 | 63 | 2.3 | 1.72 | 0.408 | 7.33 | 0.92 |
| 2 | 3.1 | 11.7 | 81.0 | 2467 | 2821 | 9108 | 84 | 2.2 | 1.62 | 0.377 | 7.72 | 0.87 |
| 3 | 3.9 | 9.5 | 82.3 | 2546 | 3300 | 11716 | 98 | 2.2 | 1.58 | 0.361 | 7.84 | 0.85 |
| 4 | 4.5 | 8.4 | 82.0 | 2566 | 3879 | 13353 | 116 | 2.2 | 1.58 | 0.345 | 8.22 | 0.82 |
| 5 | 5.7 | 6.6 | 81.3 | 2550 | 4563 | 16689 | 136 | 2.2 | 1.57 | 0.330 | 8.30 | 0.82 |
| 6 | 6.4 | 5.8 | 81.9 | 2517 | 5031 | 18967 | 150 | 2.2 | 1.58 | 0.323 | 8.75 | 0.79 |
| 7 | 7.1 | 5.6 | 80.4 | 2719 | 5559 | 20974 | 166 | 2.2 | 1.57 | 0.315 | 8.98 | 0.79 |
| 8 | 7.9 | 4.9 | 80.4 | 2648 | 5878 | 23078 | 176 | 2.2 | 1.59 | 0.310 | 9.48 | 0.75 |

#### For 0.07% of $Al_2O_3$ Nanofluid

| Sr. No. | m (kg/sec) | ΔT (°C) | $T_b$ (°C) | Q (W) | h (W/m²K) | Re | Nu | Pr | $\frac{N_{ut}}{N_{up}}$ | f | $\frac{f_t}{f_p}$ | η |
|---|---|---|---|---|---|---|---|---|---|---|---|---|
| 1 | 2.1 | 22.6 | 77.3 | 2393 | 2670 | 6106 | 66 | 1.8 | 2.16 | 0.420 | 7.63 | 1.15 |
| 2 | 2.7 | 14.9 | 82.1 | 2453 | 3308 | 8361 | 81 | 1.7 | 2.03 | 0.392 | 7.88 | 1.06 |
| 3 | 3.4 | 13.6 | 82.8 | 2549 | 4137 | 10652 | 101 | 1.7 | 1.99 | 0.372 | 8.26 | 1.03 |
| 4 | 4.2 | 10.3 | 83.4 | 2630 | 4750 | 13084 | 117 | 1.7 | 1.94 | 0.363 | 8.49 | 0.98 |
| 5 | 5.2 | 8.4 | 83.5 | 2620 | 5559 | 16247 | 136 | 1.7 | 1.90 | 0.342 | 8.67 | 0.97 |
| 6 | 5.9 | 8.1 | 82.0 | 2651 | 6470 | 18224 | 159 | 1.7 | 1.95 | 0.337 | 9.06 | 0.98 |
| 7 | 6.7 | 6.6 | 81.3 | 2673 | 6862 | 20387 | 169 | 1.7 | 1.90 | 0.330 | 9.49 | 0.91 |
| 8 | 7.6 | 6.2 | 80.9 | 2682 | 7560 | 22961 | 186 | 1.7 | 1.88 | 0.319 | 9.74 | 0.92 |

#### For 0.14% of $Al_2O_3$ Nanofluid

| Sr. No. | m (kg/sec) | ΔT (°C) | $T_b$ (°C) | Q (W) | h (W/m²K) | Re | Nu | Pr | $\frac{N_{ut}}{N_{up}}$ | f | $\frac{f_t}{f_p}$ | η |
|---|---|---|---|---|---|---|---|---|---|---|---|---|
| 1 | 1.9 | 18.0 | 84.2 | 2345 | 3400 | 5917 | 70 | 1.4 | 2.30 | 0.432 | 7.78 | 1.26 |
| 2 | 2.7 | 18.3 | 82.7 | 3250 | 4502 | 8483 | 92 | 1.4 | 2.14 | 0.397 | 8.21 | 1.19 |
| 3 | 3.2 | 13.0 | 85.5 | 2796 | 5350 | 10374 | 109 | 1.3 | 2.03 | 0.384 | 8.36 | 1.12 |
| 4 | 4.2 | 10.9 | 84.5 | 3036 | 6285 | 13259 | 128 | 1.4 | 2.02 | 0.365 | 8.69 | 1.13 |
| 5 | 5.2 | 8.63 | 84.3 | 3011 | 7356 | 16672 | 150 | 1.4 | 1.96 | 0.357 | 9.05 | 1.04 |
| 6 | 5.9 | 7.01 | 83.5 | 2739 | 8159 | 18916 | 166 | 1.3 | 1.99 | 0.343 | 9.38 | 0.98 |
| 7 | 6.4 | 5.3 | 82.7 | 2277 | 7980 | 19725 | 163 | 1.4 | 1.94 | 0.338 | 9.73 | 0.93 |
| 8 | 7.4 | 6.1 | 81.6 | 3033 | 9781 | 22850 | 200 | 1.4 | 1.92 | 0.327 | 10.20 | 0.94 |

#### For 0.21% of $Al_2O_3$ Nanofluid

| Sr. No. | m (kg/sec) | ΔT (°C) | $T_b$ (°C) | Q (W) | h (W/m²K) | Re | Nu | Pr | $\frac{N_{ut}}{N_{up}}$ | f | $\frac{f_t}{f_p}$ | η |
|---|---|---|---|---|---|---|---|---|---|---|---|---|
| 1 | 1.9 | 20.5 | 87.2 | 2538 | 4510 | 6176 | 77 | 1.1 | 2.39 | 0.432 | 8.17 | 1.37 |
| 2 | 2.6 | 14.3 | 87.9 | 2459 | 5904 | 8798 | 100 | 1.1 | 2.17 | 0.409 | 8.38 | 1.26 |
| 3 | 3.2 | 11.6 | 89.2 | 2459 | 6940 | 10927 | 118 | 1.0 | 2.14 | 0.385 | 8.71 | 1.19 |
| 4 | 4.2 | 9.1 | 88.0 | 2493 | 8084 | 14036 | 137 | 1.1 | 2.08 | 0.373 | 9.07 | 1.16 |
| 5 | 4.8 | 7.7 | 88.0 | 2460 | 9371 | 16218 | 159 | 1.1 | 2.01 | 0.365 | 9.18 | 1.11 |
| 6 | 5.6 | 6.8 | 86.0 | 2494 | 10290 | 19046 | 175 | 1.0 | 1.99 | 0.362 | 9.71 | 1.05 |
| 7 | 6.3 | 7.2 | 83.6 | 3015 | 10544 | 20531 | 180 | 1.1 | 1.92 | 0.347 | 9.99 | 0.98 |
| 8 | 7.1 | 5.3 | 83.9 | 2494 | 12142 | 22729 | 207 | 1.1 | 1.96 | 0.343 | 10.38 | 1.03 |

**Estimated Parameter of Quadruple Counter Tapes as Parallel Quadruple Swirl Flow Generators**

### For Water

| Sr. No. | m (kg/sec) | ΔT (°C) | $T_b$ (°C) | Q (W) | h (W/m²K) | Re | Nu | Pr | $\frac{N_{ut}}{N_{up}}$ | f | $\frac{f_t}{f_p}$ | η |
|---|---|---|---|---|---|---|---|---|---|---|---|---|
| 1 | 2.3 | 15.5 | 76.8 | 2407 | 2195 | 6389 | 66 | 2.3 | 1.72 | 0.418 | 7.52 | 0.97 |
| 2 | 3.1 | 11.9 | 81.0 | 2502 | 2932 | 9083 | 87 | 2.2 | 1.62 | 0.386 | 7.91 | 0.89 |
| 3 | 3.9 | 9.7 | 82.4 | 2586 | 3475 | 11793 | 103 | 2.1 | 1.58 | 0.370 | 8.04 | 0.88 |
| 4 | 4.5 | 8.3 | 82.1 | 2550 | 4001 | 13430 | 119 | 2.2 | 1.58 | 0.354 | 8.43 | 0.84 |
| 5 | 5.6 | 6.7 | 81.6 | 2566 | 4726 | 16745 | 141 | 2.2 | 1.57 | 0.338 | 8.51 | 0.84 |
| 6 | 6.4 | 5.8 | 82.8 | 2523 | 5196 | 19246 | 155 | 2.1 | 1.58 | 0.331 | 8.97 | 0.82 |
| 7 | 7.2 | 5.5 | 80.8 | 2689 | 5755 | 20875 | 172 | 2.2 | 1.57 | 0.323 | 9.20 | 0.81 |
| 8 | 7.9 | 5 | 81.2 | 2669 | 6080 | 23060 | 181 | 2.2 | 1.59 | 0.318 | 9.72 | 0.78 |

### For 0.07% of $Al_2O_3$ Nanofluid

| Sr. No. | m (kg/sec) | ΔT (°C) | $T_b$ (°C) | Q (W) | h (W/m²K) | Re | Nu | Pr | $\frac{N_{ut}}{N_{up}}$ | f | $\frac{f_t}{f_p}$ | η |
|---|---|---|---|---|---|---|---|---|---|---|---|---|
| 1 | 2.1 | 20.9 | 80.5 | 2943 | 2995 | 6519 | 73 | 1.7 | 2.30 | 0.426 | 7.82 | 1.27 |
| 2 | 2.7 | 12.6 | 84.9 | 2294 | 3557 | 8602 | 87 | 1.6 | 2.13 | 0.402 | 8.08 | 1.13 |
| 3 | 3.4 | 11.2 | 85.6 | 2551 | 4379 | 11069 | 107 | 1.6 | 2.01 | 0.378 | 8.38 | 1.08 |
| 4 | 4.2 | 8.8 | 85.4 | 2491 | 5072 | 13634 | 124 | 1.6 | 2.03 | 0.369 | 8.62 | 1.04 |
| 5 | 5.1 | 7.5 | 85.0 | 2565 | 5975 | 16497 | 146 | 1.6 | 1.93 | 0.347 | 8.89 | 1.03 |
| 6 | 5.9 | 6.5 | 83.7 | 2554 | 6629 | 19207 | 162 | 1.6 | 1.99 | 0.346 | 9.28 | 1.00 |
| 7 | 6.7 | 5.1 | 82.9 | 2300 | 6874 | 20260 | 169 | 1.7 | 1.90 | 0.331 | 9.72 | 0.91 |
| 8 | 7.6 | 5.1 | 82.2 | 2602 | 7837 | 22858 | 193 | 1.8 | 1.97 | 0.320 | 9.98 | 0.95 |

### For 0.14% of $Al_2O_3$ Nanofluid

| Sr. No. | m (kg/sec) | ΔT (°C) | $T_b$ (°C) | Q (W) | h (W/m²K) | Re | Nu | Pr | $\frac{N_{ut}}{N_{up}}$ | f | $\frac{f_t}{f_p}$ | η |
|---|---|---|---|---|---|---|---|---|---|---|---|---|
| 1 | 2.0 | 19.8 | 84.3 | 2580 | 3764 | 6170 | 77 | 1.4 | 2.54 | 0.443 | 7.98 | 1.38 |
| 2 | 2.6 | 17 | 83.8 | 2467 | 4618 | 8528 | 94 | 1.3 | 2.20 | 0.410 | 8.41 | 1.21 |
| 3 | 3.2 | 11.1 | 86.8 | 2351 | 5319 | 10663 | 108 | 1.3 | 2.11 | 0.398 | 8.57 | 1.10 |
| 4 | 4.2 | 11 | 84.7 | 2563 | 6549 | 13823 | 133 | 1.3 | 2.10 | 0.374 | 8.91 | 1.16 |
| 5 | 5.2 | 7.99 | 85.0 | 2760 | 7504 | 16946 | 153 | 1.3 | 1.99 | 0.362 | 9.19 | 1.05 |
| 6 | 5.9 | 8 | 83.0 | 2698 | 8580 | 19712 | 174 | 1.3 | 2.05 | 0.355 | 9.89 | 1.02 |
| 7 | 6.4 | 6.29 | 82.1 | 2668 | 8426 | 20037 | 172 | 1.4 | 1.95 | 0.350 | 9.98 | 0.98 |
| 8 | 7.5 | 6.69 | 81.3 | 2699 | 10328 | 23263 | 211 | 1.4 | 1.96 | 0.342 | 10.56 | 0.99 |

### For 0.21% of $Al_2O_3$ Nanofluid

| Sr. No. | m (kg/sec) | ΔT (°C) | $T_b$ (°C) | Q (W) | h (W/m²K) | Re | Nu | Pr | $\frac{N_{ut}}{N_{up}}$ | f | $\frac{f_t}{f_p}$ | η |
|---|---|---|---|---|---|---|---|---|---|---|---|---|
| 1 | 1.9 | 19.8 | 88.1 | 2466 | 4848 | 6514 | 82 | 1.0 | 2.63 | 0.452 | 8.30 | 1.46 |
| 2 | 2.6 | 14.3 | 88.2 | 2443 | 6109 | 8946 | 104 | 1.0 | 2.39 | 0.427 | 8.76 | 1.29 |
| 3 | 3.2 | 11.5 | 89.2 | 2422 | 6836 | 11107 | 116 | 1.0 | 2.19 | 0.403 | 9.10 | 1.25 |
| 4 | 4.2 | 9.07 | 88.2 | 2509 | 8481 | 14462 | 144 | 1.0 | 2.21 | 0.397 | 9.11 | 1.20 |
| 5 | 4.8 | 7.75 | 88.2 | 2442 | 9635 | 16481 | 163 | 1.0 | 2.06 | 0.371 | 9.50 | 1.13 |
| 6 | 5.6 | 6.77 | 86.2 | 2511 | 10896 | 19629 | 185 | 1.0 | 2.11 | 0.367 | 10.24 | 1.11 |
| 7 | 6.3 | 6.01 | 84.6 | 2488 | 11001 | 20629 | 187 | 1.1 | 2.00 | 0.360 | 10.34 | 1.06 |
| 8 | 7.1 | 5.34 | 84.1 | 2510 | 12929 | 23432 | 220 | 1.1 | 2.02 | 0.348 | 10.85 | 1.04 |

**Estimated Parameter of Quadruple Counter Tapes as Counter Quadruple Swirl Flow Generators**

### For Water

| Sr. No. | m (kg/sec) | $\Delta T$ ($^0C$) | $T_b$ ($^0C$) | Q (W) | h (W/m²K) | Re | Nu | Pr | $\dfrac{N_{ut}}{N_{up}}$ | f | $\dfrac{f_t}{f_p}$ | η |
|---|---|---|---|---|---|---|---|---|---|---|---|---|
| 1 | 2.3 | 15.3 | 77.5 | 2370 | 2251 | 6374 | 67 | 2.3 | 1.72 | 0.433 | 7.79 | 0.98 |
| 2 | 3.1 | 12.1 | 81.1 | 2542 | 3077 | 9075 | 92 | 2.2 | 1.62 | 0.400 | 8.20 | 0.92 |
| 3 | 3.9 | 10.2 | 82.5 | 2715 | 3698 | 11775 | 110 | 2.1 | 1.58 | 0.383 | 8.33 | 0.92 |
| 4 | 4.5 | 8.8 | 81.4 | 2684 | 4221 | 13333 | 126 | 2.2 | 1.58 | 0.366 | 8.73 | 0.87 |
| 5 | 5.7 | 6.7 | 82.5 | 2587 | 4896 | 16877 | 146 | 2.2 | 1.57 | 0.351 | 8.82 | 0.86 |
| 6 | 6.4 | 6 | 83.0 | 2588 | 5474 | 19083 | 163 | 2.1 | 1.58 | 0.343 | 9.30 | 0.85 |
| 7 | 7.1 | 5.7 | 80.2 | 2759 | 5962 | 20669 | 178 | 2.2 | 1.57 | 0.335 | 9.54 | 0.83 |
| 8 | 8.0 | 4.9 | 81.5 | 2649 | 6258 | 23356 | 187 | 2.2 | 1.59 | 0.329 | 10.07 | 0.80 |

### For 0.07% of $Al_2O_3$ Nanofluid

| Sr. No. | m (kg/sec) | $\Delta T$ ($^0C$) | $T_b$ ($^0C$) | Q (W) | h (W/m²K) | Re | Nu | Pr | $\dfrac{N_{ut}}{N_{up}}$ | f | $\dfrac{f_t}{f_p}$ | η |
|---|---|---|---|---|---|---|---|---|---|---|---|---|
| 1 | 2.1 | 25.5 | 76.9 | 2928 | 3022 | 6162 | 75 | 1.8 | 2.55 | 0.446 | 8.11 | 1.28 |
| 2 | 2.7 | 15.9 | 82.6 | 2346 | 3702 | 8333 | 91 | 1.7 | 2.33 | 0.421 | 8.38 | 1.22 |
| 3 | 3.5 | 14.5 | 83.4 | 2687 | 4697 | 10606 | 115 | 1.7 | 2.20 | 0.396 | 8.69 | 1.15 |
| 4 | 4.3 | 11.0 | 83.7 | 2552 | 5387 | 13212 | 132 | 1.7 | 2.20 | 0.386 | 9.02 | 1.09 |
| 5 | 5.2 | 9.6 | 83.4 | 2686 | 6347 | 16172 | 156 | 1.7 | 2.09 | 0.359 | 9.21 | 1.08 |
| 6 | 6.0 | 7.5 | 82.8 | 2434 | 6860 | 18404 | 168 | 1.7 | 2.15 | 0.355 | 9.62 | 1.02 |
| 7 | 6.8 | 6.5 | 81.7 | 2364 | 7315 | 20561 | 180 | 1.7 | 2.02 | 0.354 | 10.18 | 0.96 |
| 8 | 7.7 | 6.2 | 81.2 | 2564 | 8154 | 23198 | 201 | 1.8 | 2.03 | 0.338 | 10.34 | 0.97 |

### For 0.14% of $Al_2O_3$ Nanofluid

| Sr. No. | m (kg/sec) | $\Delta T$ ($^0C$) | $T_b$ ($^0C$) | Q (W) | h (W/m²K) | Re | Nu | Pr | $\dfrac{N_{ut}}{N_{up}}$ | f | $\dfrac{f_t}{f_p}$ | η |
|---|---|---|---|---|---|---|---|---|---|---|---|---|
| 1 | 2.0 | 18.1 | 85.7 | 2396 | 4019 | 6044 | 82 | 1.4 | 2.72 | 0.460 | 8.51 | 1.46 |
| 2 | 2.7 | 18.8 | 83.7 | 3355 | 5103 | 8552 | 104 | 1.4 | 2.50 | 0.421 | 8.72 | 1.32 |
| 3 | 3.2 | 9.2 | 88.6 | 1980 | 5939 | 10452 | 121 | 1.3 | 2.36 | 0.408 | 8.88 | 1.29 |
| 4 | 4.3 | 10.9 | 85.4 | 3077 | 7280 | 13544 | 148 | 1.4 | 2.34 | 0.395 | 9.23 | 1.28 |
| 5 | 5.3 | 7.1 | 86.0 | 2507 | 8140 | 16793 | 166 | 1.4 | 2.17 | 0.375 | 9.62 | 1.13 |
| 6 | 6.0 | 8.4 | 83.2 | 3349 | 9487 | 19321 | 193 | 1.3 | 2.20 | 0.368 | 10.26 | 1.12 |
| 7 | 6.5 | 5.9 | 82.8 | 2539 | 9237 | 19874 | 189 | 1.4 | 2.04 | 0.367 | 10.34 | 1.06 |
| 8 | 7.6 | 5.2 | 82.4 | 2638 | 10246 | 23341 | 209 | 1.4 | 2.07 | 0.355 | 10.94 | 0.97 |

### For 0.21% of $Al_2O_3$ Nanofluid

| Sr. No. | m (kg/sec) | $\Delta T$ ($^0C$) | $T_b$ ($^0C$) | Q (W) | h (W/m²K) | Re | Nu | Pr | $\dfrac{N_{ut}}{N_{up}}$ | f | $\dfrac{f_t}{f_p}$ | η |
|---|---|---|---|---|---|---|---|---|---|---|---|---|
| 1 | 1.9 | 16.6 | 90.6 | 2094 | 5909 | 6611 | 100 | 1.0 | 2.98 | 0.464 | 8.93 | 1.64 |
| 2 | 2.6 | 15.9 | 88.0 | 2763 | 6704 | 9079 | 114 | 1.0 | 2.56 | 0.443 | 8.91 | 1.46 |
| 3 | 3.2 | 10.2 | 90.6 | 2168 | 8046 | 11409 | 136 | 1.0 | 2.47 | 0.425 | 9.34 | 1.45 |
| 4 | 4.3 | 9.5 | 88.4 | 2657 | 9569 | 14677 | 162 | 1.0 | 2.39 | 0.396 | 9.82 | 1.34 |
| 5 | 4.9 | 7.7 | 88.5 | 2478 | 10614 | 16726 | 180 | 1.0 | 2.27 | 0.392 | 9.94 | 1.23 |
| 6 | 5.7 | 3.4 | 88.2 | 1291 | 12004 | 20162 | 203 | 1.0 | 2.25 | 0.380 | 10.51 | 1.20 |
| 7 | 6.4 | 5.9 | 84.9 | 2479 | 11835 | 21191 | 201 | 1.1 | 2.04 | 0.380 | 10.82 | 1.13 |
| 8 | 7.3 | 5.3 | 84.1 | 2547 | 13010 | 23781 | 221 | 1.1 | 2.09 | 0.361 | 10.91 | 1.04 |

# PUBLICATIONS BASED ON PRESENT WORK

## Sample Calculation

Sample calculations for Quadruple Counter Tapes as Counter Quadruple Swirl Flow Generators with water as a base fluid.

**Measured Data:**

| Parameter | Notation | Value |
|---|---|---|
| Time for flow of 1litre of fluid (sec) | t | 26.36 |
| Inlet Temp of Fluid (0C) | Tin | 69.8 |
| Outlet Temp of Fluid (0C) | Tout | 85.1 |
| Average surface Temp (0C) | Ts | 94.20 |
| Pressure Drop (Pa) | ΔP | 153.6 |

1. Mass flow rate and velocity:

$$\dot{Q} = \frac{1}{t} = \frac{1}{26.36} = 0.0379 \frac{l}{sec} = \frac{0.0379}{1000} = 0.0000379 \frac{m^3}{sec}$$

$$\dot{m} = \frac{(\rho)(\dot{Q})}{10^3} = \frac{(973.8)(0.0379)}{10^3} = 0.0369 \frac{kg}{sec}$$

$$v = \frac{\dot{Q}}{A} = \frac{0.0000379}{\left(\frac{\pi}{4}\right)(0.02)^2} = 0.121 \frac{m}{sec}$$

2. Temperature Difference and Bulk Mean Temperature

$$\Delta T = T_{out} - T_{in} = 85.1 - 69.8 = 15.3 \, °C$$

$$T_b = \frac{T_{out} + T_{in}}{2} = \frac{85.1 + 69.8}{2} = 77.45 \, °C$$

Thermal properties of water is selected at $T_b = 77 \, °C$

| Property | Notation | Value |
|---|---|---|
| Dynamic Viscosity (N-s/m2) | µ | 0.36888 X 10-3 |
| Density(kg/m3) | ϱ | 973.8 |
| Specific Heat (J/kg K) | Cp | 4193.8 |
| Thermal Conductivity ( W/mK) | k | 0.668 |

3. Reynolds Number

$$R_e = \frac{\rho v D}{\mu} = \frac{(973.8)(0.121)(0.02)}{0.36888 \times 10^{-3}} = 6374$$

4. Rate of Heat Transfer

$$Q = mC_p \Delta T = (0.0369)(4193.8)(15.3) = 2370.4 \text{ W}$$

5. Overall heat transfer coefficient

$$h = \frac{Q}{(A)(\Delta T)} = \frac{2370.4}{(\pi)(0.02)(1)(15.3)} = 2251.4 \frac{W}{m^2 K}$$

6. Nusselt number and Nusselt Number Ratio

$$N_u = \frac{hD}{k} = \frac{(2251.4)(0.02)}{0.668} = 67.41$$

$$\frac{N_{ut}}{N_{up}} = 1.72$$

7. Prandtl number

$$P_r = \frac{\mu C_p}{k} = \frac{(0.36888 \times 10^{-3})(4193.8)}{0.668} = 2.32$$

8. Friction factor and Friction Factor Ratio

$$f = \frac{(2)(D)(\nabla P)}{(L)(\rho)(v^2)} = \frac{(2)(0.02)(153.6)}{(1)(973.8)(0.121)} = 0.43302$$

$$\frac{f_t}{f_p} = 7.79$$

9. Thermal performance factor

$$\frac{\frac{N_{ut}}{N_{up}}}{\left(\frac{f_t}{f_p}\right)^{1/3}} = \frac{1.72}{7.79^{1/3}} = 0.98$$

# ABOUT THE AUTHOR

Prof. Md. Sadique Chakole is currently working as an Assistant Professor at G.H. Raisoni College of Engineering and Management, Ahmednagar, which is one of the best reputed institutes in India. He has completed his master degree with distinction and special awards from the institute. He has completed his bachelor degree with distinction in each semester and he has secured the first rank for a number of times. His project was awarded as best project in various national and international competitions. He has won more than ten prices at national events for his paper presentations.